Basics of Space Flight

BLACK & WHITE EDITION

Dave Doody

BLÜROOF PRESS

Cover illustrations:

The background is an infrared view, at 3.6 and 4.5 microns wavelength, of a colony of hot, young stars in the Orion nebula, from NASA's Spitzer Space Telescope. Arbitrary shades represent the invisible infrared wavelengths. Image (PIA-13005) courtesy NASA/JPL-Caltech.

The rocket is a Titan IVB/Centaur carrying the Cassini Saturn orbiter and its attached Huygens Titan probe. Launch occurred at 4:43 a.m. EDT October 15, 1997, from Cape Canaveral Air Station. Image (PIA-01050) courtesy NASA/JPL/KSC.

Inserts to the lower left of rocket: Leftmost is a view of the Sun using data from the Solar & Heliospheric Observatory spacecraft, displayed by the JHelioviewer visualization software (www.jhelioviewer.org). Center image combines Magellan radar imaging and altimetry data in a view of volcanoes "Innini" and "Hathor" in the southern hemisphere of Venus. From image (P-42385) courtesy of the Magellan Science Team at the Jet Propulsion Laboratory Multimission Image Processing Laboratory. Rightmost: NASA/JPL 70-meter Deep Space Station at Goldstone, California. View looking toward the northeast. Courtesy NASA/JPL-Caltech.

Basics of Space Flight

Black & White POD Edition, JPL Clearance #11-1594

Prepared with LaTeX from public-domain content at www.jpl.nasa.gov/basics. Subject content is in the public domain and may be copied without permission. Please acknowledge source as the above website when copying. The work was carried out at the Jet Propulsion Laboratory, California Institute of Technology, under a contract with the National Aeronautics and Space Administration. Free-downloadable PDF version of this book is available at the above URL.

Published in the United States by BLüROOF PRESS,
an activity of SCI / Space Craft International, Pasadena.

ISBN-13: 978-0615484112

ISBN-10: 0615484115

9 8 7 6 5 4 3 2 1

Contents

List of Figures

List of Tables

The Black & White Edition

This edition is printed in black and white in order to make low-cost copies widely available. A full-color edition of *Basics of Space Flight* (ISBN 978-0615476018) is also in print, and may be found wherever books are sold.

2

Author's Preface

In January of this year, I published a little book using an old manuscript of my late father's. My purpose was to get a handle on this new system of "POD" — Print On Demand, which includes publishing, marketing, selling and shipping, all at no cost to the author. I did this just to see how easy or difficult a POD task might be, with the ultimate intention of perhaps some day creating the present book the same way. The "test book" was easy.

Then, after it was done, I searched `amazon.com`, and saw it, of course, along with the book I wrote a couple years ago, *Deep Space Craft* (Springer, 2009). The latter, by the way, does not ignore all the math, the way this one, which is transcribed from the JPL public website, blatantly does.

But there among them on Amazon was, to my great surprise, another entry under my name. It said, "No image available" and listed *Basics of Space Flight* (JPL 1993) for a ridiculously high price. Amazon or one of its affiliates had somehow gotten an old spiral-bound hardcopy of the original training workbook that I co-authored with George Stephan some eighteen years ago (before the Web as we know it existed). Perfectly legal, of course, as it's all in the public domain. But since 1993, JPL's *Basics of Space Flight* website* (which I try to keep updated while working full-time on *Cassini*) has expanded far beyond that ancient manual. This motivated me to get moving to create a more up-to-date and higher quality hardcopy. The result is this 2011 snapshot of the JPL *Basics* site made via Print On Demand. This hardcopy book was entirely an after-hours project, as was most of the online tutorial to begin with.

Why go to the trouble? I love to learn and teach about how things work. And I think robotic space flight generally enjoys enormously less public attention than it deserves.

In my years as an engineer in robotic space flight operations, I've been fortunate to enjoy front-row seating to a most *amazing* human endeavor.

---

*http://www.jpl.nasa.gov/basics

After centuries upon centuries of dreaming about it, today we are actually exploring our cosmic environs with robot rocket ships, while the usual media outlets pay little attention. The worlds in our back yard are absolutely spectacular to encounter, and they're rich with information about how planets and life evolve while orbiting a star. This is compelling stuff. It has flourished because of international cooperation and human understanding, and it's too bright a light to be overlooked amid everything else that's important in our world.

As if all this weren't enough, the finest book production tools are right at hand, waiting for anybody to use them, free of charge, thanks to the current-day Ben Franklins of the world.

Dave Doody
Altadena, California
April 20, 2011

Acknowledgements

The present book is a reformatted copy of public-domain content from the online *Basics of Space Flight* JPL public website. It has been paginated and modified to treat selected web links as notes, and images as referenced figures. Color illustrations have been rendered in grayscale. The index was newly created to accommodate the loss of electronic search capability that the online site offers. Images are courtesy NASA/JPL-Caltech unless otherwise specified. Thanks to Greg Chin, Arden Acord, and Victoria Ryan for reviewing this edition, to Adam Cochran for the intellectual property guidance, and to Adria Roode for shepherding the project through the system.

A major "21st Century Edition" website re-write* was undertaken starting in 2000 by Dave Doody and Diane Fisher, working under the auspices of the JPL Mission Execution & Automation Section. Diane created most of the animated images on the website, and performed technical editing of the entire document. The online version lists contributors who helped with various facets of the electronic site such as animation and quiz-scoring mechanisms, but of course those are not part of this book.

Thanks to Susan Watanabe and Mary Beth Murrill of Media Relations for their reviews and advice pertaining to the online 21st Century Edition. Thanks also to Susan Kurtik for funding that work, to Ben Toyoshima for guidance, to Bill Kurth for help with information about the heliosphere, to Jeremy Jones for reviewing navigation issues, and to Troy Goodson and Duane Roth for their illuminating input on navigation. Thanks to Steve Edberg for help with things astronomical, Laura Sakamoto, Steve Slobin, and Robert Sniffin for expertise with telecommunications and DSN topics, to Betsy Wilson for lots of help with the telemetry section, and to Trina Ray for enthusiastically reviewing Radio Science and DSN topics. Thanks to Greg Chin, manager of the Cassini Mission Support & Services Office, who granted the author freedom to work on the online 21st Century edition

---

*JPL Document D-20120, Review Clearance CL-03-0371.

5

during a busy Jupiter flyby period for the Cassini Mission to Saturn, and to Bob Mitchell for allowing time to keep it current as the Cassini Saturn Tour and extended missions progress.

For the original 1993 version co-authored with George Stephan: Diane Fisher provided technical editing and illustration, and took the initiative to publish it on the fledgling world-wide web. Cozette Parker assisted with the initial hardcopy publication. Brad Compton kindly tolerated the author's preoccupation with this project during Magellan's demanding mission at Venus. Special thanks to the original reviewers Ben Toyoshima, Larry Palkovic, Carol Scott, Rob Smith, Dan Lyons, and Bob Molloy, and to field testers Kathy Golden, Linda Lee, Paul Porter, and the late Steve Annan for their valuable comments. Thanks to Roy Bishop (Physics Department, Acadia University, and the Royal Astronomical Society of Canada) for his independent review of that early work.

From 1995 through 1997, The Planetary Society published my regular column, "Basics of Space Flight," which drew on and embellished the original tutorial. Some of those articles still appear in the "publications" section on The Planetary Society's website, and aspects of them were incorporated into the 21st Century Edition on the web with their kind permission.

I am indebted to many readers of the online site who have pointed out various needed corrections over the years.

In memory of Donald E. Whiting (1946-1993).

Introduction

The people of Caltech's Jet Propulsion Laboratory create, manage, and operate NASA projects of exploration throughout our solar system and beyond.

Basics of Space Flight has evolved from a tutorial that was designed in 1993 to help operations people identify the range of concepts associated with deep space missions, and to grasp the relationships among them. It became a website* as the web itself evolved, and as such it has gained popularity with college and high-school students and faculty, and people everywhere who want to know more about interplanetary space flight.

This tutorial attempts to offer a broad scope, but limited depth, serving as a robust framework to accommodate further training or investigation. Many other resources are available for delving into each of the topics related here; indeed, any one of them can involve a lifelong career of specialization. This book's purpose is met if the reader learns the scope of concepts that apply to interplanetary space exploration, and how the relationships among all of them work.

Except for units of measure, abbreviated terms are spelled out the first time they are used. Thereafter, the abbreviations are generally used alone, and they can be found in the Glossary. Throughout this tutorial, measurements are expressed in SI, the International System of Units (the modern metric system of measurement). You will find them explained, along with a table of metric to English conversions, under Units of Measure starting on page 269.

---

*http://www.jpl.nasa.gov/basics

Part I

ENVIRONMENT

Chapter 1

The Solar System

> **Objectives:** Upon completion of this chapter, you will be able to classify objects within the solar system, state their distances of in terms of light-time, describe the Sun as a typical star, relate its share of the mass within the solar system, and compare the terrestrial and Jovian planets. You will be able to distinguish between inferior and superior planets, classical and dwarf planets, describe asteroids, comets, the Kuiper belt, and the Oort cloud. You will be able to describe the environment in which the solar system resides.

The solar system has been a topic of study from the beginning of history. For nearly all that time, people have had to rely on long-range and indirect measurements of its objects. For all of human history and pre-history, observations were based on visible light. Then in the twentieth century, people discovered how to use additional parts of the spectrum. Radio waves, received here on Earth, have been used since 1931[1] to investigate celestial objects. Starting with the emergence of space flight in 1957,[2] instruments operating above Earth's obscur-

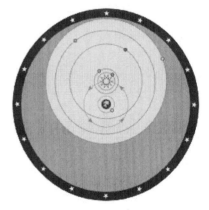

Figure 1.1: Geocentric system of Tycho Brahe (1546-1601). Image courtesy Wikimedia Commons.

ing atmosphere could take advantage not only of light and radio, but virtually the whole spectrum (the electromagnetic spectrum is the subject of a later chapter). At last, with interplanetary travel, instruments can be carried to many solar system objects, to measure their physical properties and dynamics directly and at very close range. In the twenty-first century,

13

knowledge of the solar system is advancing at an unprecedented rate.

The solar system consists of an average star we call the Sun, its "bubble" the heliosphere, which is made of the particles and magnetic field emanating from the Sun—the interplanetary medium— and objects that orbit the Sun: from as close as the planet Mercury all the way out to comets almost a light-year away.

Figure 1.1 shows the geocentric system envisioned by Tycho Brahe (1546-1601) with the Moon and the Sun revolving around Earth while Mercury, Venus, Mars, Jupiter, and Saturn revolve around the Sun, all surrounded by a sphere of fixed stars. Tycho's assistant Johannes Kepler later based his refinement of Nicolaus Copernicus's heliocentric system upon Tycho's lifetime compilation of astronomical data.

Objects that orbit the Sun are given many names and classifications, some of which overlap. The planets Mercury, Venus, Earth, Mars, Jupiter, and Saturn were known to the ancients (though Earth itself may not have been recognized as a planet until Nicolaus Copernicus's publication in 1543). The planets Uranus and Neptune, visible only with telescopic aid, weren't discovered until 1781 and 1846, respectively. Many of the planets have their own retinue of diverse satellites or moons.

In 1801, when an object was discovered orbiting the Sun between Mars and Jupiter, it was heralded as a new planet. Shortly thereafter, though, many more such objects were discovered in the same region, so the classification "asteroid" or "minor planet" became appropriate. There are many classes of asteroids, as you will see later in this chapter.

The four inner planets are known to be rocky, and they are classified as terrestrial planets. The four outer planets Jupiter, Saturn, Uranus, and Neptune, are gas giants —largely hydrogen and some helium, with small solid cores. They're also called Jovian planets.

In 1930, an object was discovered orbiting the Sun beyond Neptune; it was named Pluto. Its highly elliptical orbit was found to be inclined an unprecedented 17 degrees from the ecliptic. Like the first asteroid, it was heralded as a new planet. Fast-forward to the age of powerful new telescopes, and we learn that Pluto was only the first of many such objects to be discovered in that region of space. So, in 2006, the International Astronomical Union undertook to define the term planet for the first time.[3] Pluto was redefined as one of five dwarf planets, another of which is Ceres, the first-seen asteroid.

Of the objects orbiting the Sun beyond Neptune, called Trans-Neptunian Objects, TNO, Pluto and the three other other dwarf planets (known as of early 2011) in that realm are called plutoids. They are: Haumea, Makemake, and Eris. These, and many other bodies, are members of the vast Kuiper Belt. Composed of material left over after the formation of the other planets (see Comets, page 40), Kuiper Belt Objects (KBO) were never exposed to the higher temperatures and solar radiation levels of the inner solar system. They remain as a sample of the primordial material that set the stage for the evolution of the solar system as it exists today, including life. The New Horizons Spacecraft is en route for a 2015 flyby of Pluto, to examine at least this one KBO.

The discovery of a TNO known as Sedna, which is smaller than Pluto, was of great scientific interest because of its very distant and highly elongated orbit. The closest it comes to the Sun, 76.4 AU (an astronomical unit, AU, represents the distance between the Earth and Sun; see page 18) is more than twice Neptune's distance. Sedna is currently outbound in its 12,000-year orbit to a high point of 961 AU from the Sun. Understanding the genesis of its unusual orbit is likely to yield valuable information about the origin and early evolution of the Solar System. Some astronomers consider that it may be the first known member of the Oort cloud.

The Oort cloud is a hypothesized spherical reservoir of comet nuclei orbiting roughly 50,000 AU, nearly a light-year, from the Sun. No confirmed direct observations of the Oort cloud have been made yet, but astronomers believe that it is the source of all long-period comets that enter the inner Solar System when their orbits are disturbed.

In Cosmic Perspective

Interstellar space is the term given to the space between stars within the galaxy. The Sun's nearest known stellar neighbor is a red dwarf star called Proxima Centauri, at a distance of about 4.2 light years (a light year is the distance light travels in a year, at about 300,000 km per second). We are beginning to find that many stars besides the Sun harbor their own "solar systems" with planets, which are being called extrasolar planets, or exoplanets. As of early February 2011, astronomers have detected a total of 526 planets[4] orbiting other stars. This number is up from the milestone of 100 exoplanets discovered as of January 2004. Most known exoplanets are gas giants, Jupiter-like planets, since current methods favor the detection of the more massive worlds. Most are relatively nearby, within 5,000 light years, although one candidate was discovered in September 2005 via "microlensing" at a distance of 17,000 light years.

Our whole solar system, along with all the local stars you can see on a clear dark night, reside in one of our galaxy's spiral arms, known as the Orion arm, as they orbit the supermassive black hole in the dense star cluster at the center of our galaxy some 26,000 (±1400) light-years distant from us. At this great distance, we go around once about every 250 million years. Stars very close to the central mass are observed in keplerian orbits with periods as short as 15.2 years. This spiral disk that we call the Milky Way includes some 200 billion stars, thousands of gigantic clouds of gas and dust, and enormous quantities of mysterious dark matter.

Figure 1.2 shows a galaxy known as M100 that is similar to our own Milky Way galaxy.[5] The Milky Way has two small galaxies orbiting it nearby, which are visible from Earth's southern hemisphere. They are called the Large Magellanic Cloud and the Small Magellanic Cloud.

Our galaxy, one of billions of galaxies known, is travelling through intergalactic space. On a cosmic scale, all galaxies are generally receding from each other, although those relatively close together may exhibit additional local motion toward or away from each other as well.

Aside from its galactic orbital velocity (250-300 km/second), the Sun and its planetary system wander through the local stellar neighborhood at roughly 100,000 kilometers per hour, entering and leaving various tenuous local clouds of gas on a time scale of roughly once every few thousand to millions of years.

Immediately surrounding our solar system is a warm, partly ionized cloud, called the Local Interstellar Cloud. Like most interstellar clouds, its gas comprises about 90% hydrogen and 10% helium. Roughly 1% of the cloud's mass is dust.

Figure 1.2: Galaxy M100, one of the brightest members of the Virgo Cluster of galaxies, can be seen through a moderate-sized amateur telescope. M100 is spiral shaped, like our Milky Way, and tilted nearly face-on as seen from Earth. The galaxy has two prominent arms of bright stars and several fainter arms. Imaged by the Hubble Space Telescope.

Motions Within the Solar System

The Sun and planets each rotate on their axes. Because they formed from the same rotating disk, the planets, most of their satellites, and the asteroids, all revolve around the Sun in the same direction as the Sun rotates, and in

Table 1.1: Approximate Typical Conditions Within Our Galaxy*

Region of Interstellar Space Within our Galaxy	Number Density Atoms / cm^3	Temperature† Kelvins
Inside our Heliosphere, in the Vicinity of Earth	5	10,000
Local Cloud Surrounding our Heliosphere	0.3	7,000
Nearby Void (Local Bubble)	< 0.001	1,000,000
Typical Star-Forming Cloud	>1,000	100
Best Laboratory Vacuum	1,000	
Classroom Atmosphere	2.7 $\times 10^{19}$	288

*Outside the galaxy, in intergalactic space, the number density of particles is thought to fall off to about one atom or molecule per cubic meter $(10^{-6}$ / cm$^3)$.

†Temperatures refer to particles in the near-vacuum of space. Large physical masses, such as spacecraft, asteroids, etc., are not affected by the particles' temperature. Instead, their thermal states are dominated by solar (or stellar) heating and exposure to shadow.

nearly circular orbits. The planets orbit the Sun in or near the same plane, called the ecliptic (because it is where eclipses occur). Originally regarded as the ninth planet, Pluto seemed a special case in that its orbit is highly inclined (17 degrees) and highly elliptical. Today we recognize it as a Kuiper Belt Object. Because its orbit is so eccentric, Pluto sometimes comes closer to the Sun than does Neptune. Most planets rotate in or near the plane in which they orbit the Sun, again because they formed, rotating, out of the same dust ring. The exception, Uranus, must have suffered a whopping collision that set it rotating on its side.

Distances Within the Solar System

The most commonly used unit of measurement for distances within the solar system is the astronomical unit (AU). The AU is based on the mean distance from the Sun to Earth, roughly 150,000,000 km. JPL's Deep Space Network refined the precise value of the AU in the 1960s by obtaining radar echoes from Venus. This measurement was important since spacecraft navigation depends on accurate knowledge of the AU. Another way to indicate distances within the solar system is terms of light time, which is the distance light travels in a unit of time. Distances within the solar system, while vast compared to our travels on Earth's surface, are comparatively small-scale in astronomical terms. For reference, Proxima Centauri, the nearest star at about 4.2 light years away, is about 265,000 AU from the Sun.

Table 1.2: Light time and Distance

Light Time	Approximate Distance	Example
3 seconds	900,000 km	Earth-Moon Round Trip
3 minutes	54,000,000 km	Sun to Mercury
8.3 minutes	149,600,000 km	Sun to Earth (1 AU)
1 hour	1,000,000,000 km	1.5 x Sun-Jupiter Distance
16.1 hours	115 AU	Voyager-1 (November, 2010)
1 year	63,000 AU	Light Year
4.2 years	265,000 AU	Next closest star

Temperatures Within the Solar System

The temperature of planets and other objects in the solar system is generally higher near the Sun and colder as you move toward the outer reaches of the solar system. The temperature of low-density plasma (charged particles in the environment), though, is typically high, in the range of thousands of degrees (see "Our Bubble of Interplanetary Space" on page 23).

Table 1.3 shows examples and compares temperatures of objects and conditions from absolute zero through planet temperatures, to those of stars and beyond. It also includes temperature conversion factors.

The Sun

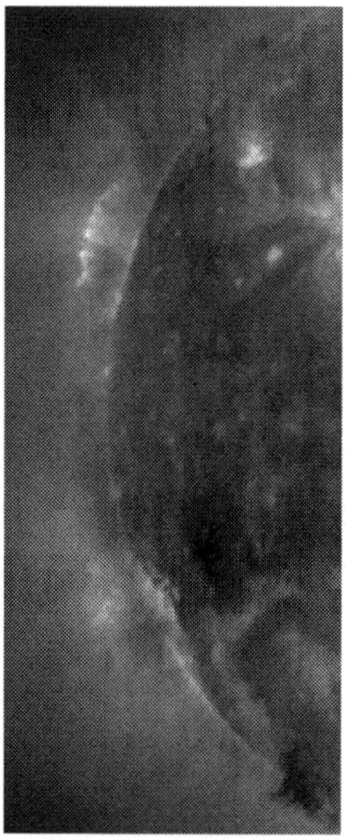

The Sun is a typical star. Its spectral classification is "G2 V." The G2 part basically means it's a yellow-white star, and the roman numeral V means it's a "main sequence" dwarf star (by far the most common) as opposed to supergiant, or sub-dwarf, etc. The fact that the Sun is the dominant source of energy for processes on Earth that sustain us makes it a subject of major interest for study. As viewed from the Earth's surface, the Sun subtends roughly half a degree of arc upon the sky (as does the Moon, at this period in the solar system's history.)

You can view some remarkable current images of the Sun as seen today by the suite of instruments aboard the SOHO (Solar & Heliospheric Observatory) spacecraft as it views the Sun from the L1 Lagrange point between the Earth and the Sun. The image of the Sun in Figure 1.3 is from a SOHO movie combining three wavelengths of extreme ultravilot light.[6]

Figure 1.3: The Sun, extreme ultravilot view. Imaged by Solar & Heliospheric Observatory.

The STEREO pair of spacecraft provide the means to study the Sun in three dimensions. One spacecraft moves ahead of the Earth in solar orbit, and is named STEREO-A; the other, STEREO-B, lags behind Earth. The latest images from that mission can be found online.[7]

Mass: Because of its enormous mass, the Sun dominates the gravita-

Table 1.3: Solar System Temperature Reference

Melting and boiling points shown to precision for pressure of 1 atmosphere. Values for stars, planet cloudtops, surfaces etc. shown as round numbers rather than precise conversions.

Kelvin	Degrees C	Degrees F	Remarks
0	−273.15	−459.67	Absolute Zero
picokelvins	−273.15∼	−459.67∼	Lowest achieved in a lab
2.7	−270.5	−454.8	Cosmic microwave background
4.2	−268.95	−452.11	Liquid helium boils
14.01	−259.14	−434.45	Solid hydrogen melts
20.28	−252.87	−423.16	Liquid hydrogen boils
35	−235	−390	Neptune's moon Triton surface
63.17	−209.98	−345.96	Solid nitrogen melts
72	−201	−330	Neptune 1-bar level
76	−197	−323	Uranus 1-bar level
77.36	−195.79	−320.42	Liquid nitrogen boils
90	−180	−300	Saturn's moon Titan surface
90.188	−182.96	−297.33	Liquid oxygen boils
100	−175	−280	Planet Mercury surface, night
134	−139	−219	Saturn 1-bar level
153	−120	−184	Mars surface, nighttime low
165	−108	−163	Jupiter 1-bar level
195	−78.15	−108.67	Carbon dioxide freezes (dry ice)
273.15	**0.0**	**32.0**	**Water ice melts**
288	15	59	Mars surface, daytime high
288.15	15.0	59.0	Standard room temperature
373.15	**100**	**212**	**Liquid water boils**
600.46	327.31	621.16	Lead melts
635	362	683	Venus surface
700	425	800	Planet Mercury surface, day
750	475	890	Uranus hydrogen "corona"
1,337.58	1,064.43	1,947.97	Solid gold melts
1,700	1,400	2,600	Candle flame, yellow mid part
3,500	3,200	5,800	Betelgeuse (red giant)
3,700	3,400	6,700	Sunspots
5,700	5,400	9,800	Solar photosphere
7,000	7,000	12,000	Typical neon sign plasma
8,000	8,000	14,000	Est. local interstellar medium
10,000	10,000	18,000	Sirius (blue-white star)
15,000	15,000	27,000	Saturn core
30,000	30,000	54,000	Jupiter core
	100,000 to 2,000,000	180,000 to 3,600,000	Heliosphere plasma
	15,000,000	27,000,000	Solar core
	100,000,000,000	(100 billion; 10^{11})	Supernova
	4,000,000,000,000	(4 trillion; 4×10^{12})	Quark-gluon plasma

$$K = {}^\circ C + 273.15 \qquad {}^\circ C = (5/9) \times ({}^\circ F - 32) \qquad {}^\circ F = (9/5) \times {}^\circ C + 32$$

tional field of the solar system. The motion of everything within a few light years of the Sun is dominated by the effect of the solar mass. At 1.98892 $\times 10^{30}$ kilograms, or roughly 333,000 times the mass of the Earth, it contains over 99 percent of the solar system's mass. The planets, which condensed out of the same disk of material that formed the Sun, contain just over a tenth of a percent the mass of the solar system.

The Sun's mass can be easily calculated.[8]

Table 1.4: Mass Distribution Within the Solar System

99.85%	Sun
0.135%	Planets
0.015%	Comets, Kuiper belt objects, Satellites of the planets, Asteroids (minor planets), Meteroids, and Interplanetary Medium

Even though the planets make up only a small portion of the solar system's mass, they do retain the vast majority of the solar system's angular momentum. This storehouse of momentum can be utilized by interplanetary spacecraft on so-called "gravity-assist" trajectories.

Fusion: The Sun's gravity creates extreme pressures and temperatures within its core, sustaining a thermonuclear reaction fusing hydrogen nuclei and producing helium nuclei. This reaction converts about 4 billion kilograms of mass to energy every second. This yields tremendous amounts of energy, causing the state of all the Sun's material to be plasma and gas. These thermonuclear reactions began about 5 billion years ago in the Sun, and will probably continue for another 5 billion years into the future. Energy produced in the core takes over a million years to reach the surface and be radiated as light and heat (view illustrated description online[9]). Our star's output varies slightly over an 11-year cycle, during which the number of sunspots changes.

Solar sound waves are mostly trapped inside the Sun. They refract away from the hot core, and reflect back and forth between different parts of the photosphere. By monitoring the Sun's vibrating surface, helioseismologists can probe the stellar interior in much the same way that geologists use seismic waves from earthquakes to probe the inside of our planet.

Rotation: The Sun rotates on its axis with a period of approximately 25.4 days. This is the adopted value at 16° latitude. Because the Sun is a gaseous body, not all its material rotates together. The Goddard solar fact sheet[10] describes how rotation varies with latitude. Solar matter at very high latitudes takes over 30 days to complete a rotation.

The Sun's axis is tilted about 7.25 degrees to the plane of the Earth's orbit, so we see a little more of the Sun's northern polar region each September and more of its southern region in March.

Magnetic Field: Magnetism is produced in the Sun by the flow of electrically charged particles: ions and electrons. Sunspots are somewhat cooler places seen on the photosphere (the Sun's bright surface) where very intense magnetic lines of force break through. Prominences that seem to float above the photosphere are supported by, and threaded through with, magnetic fields. All the streamers and loops seen in the corona (the Sun's extended upper atmosphere) are shaped by magnetic fields. Magnetic fields are at the root of virtually every feature observed on and above the Sun.

Not surprising, the Sun's magnetic field permeates interplanetary space. The magnetic field of the Sun influences the way in which charged particles (cosmic rays, solar energetic particles, and even interstellar dust grains) move through the heliosphere.

The 11-year solar magnetic cycle that sunspot activity follows, is part of a 22-year "Hale" cycle. During the first half of the 22-year cycle, the Sun's magnetic north pole is in the northern hemisphere, while the magnetic south pole is in the southern hemisphere. Right around the peak of the sunspot cycle (solar maximum) about 11 years into the magnetic cycle, the magnetic poles flip, exchanging places, so that magnetic north is then located in the southern hemisphere.

Mass Ejections: Coronal mass ejections, CMEs, are huge magnetic bubbles of plasma that expand away from the Sun at speeds as high as 2000 km per second. A single CME can carry up to ten billion tons (10^{13} kilograms) of plasma away from the Sun. Coronal mass ejections were once thought to be initiated by solar flares. Although some are accompanied by flares, it is now known that most CMEs are not associated with flares.

When a CME arrives at Earth, it can cause fluctuations in the Earth's magnetic field that can play havoc with the civil electrical power distribution infrastructure, by inducing unwanted voltages.

A spectacular 5-Mbyte SOHO movie of CMEs in August, 1999 can be seen online.[11] Stars are seen moving behind the Sun as the view from SOHO changes while Earth orbits toward the right.

Solar Wind: The solar wind streams off of the Sun in all directions. The source of the solar wind is the Sun's hot corona, whose temperature is so high that the Sun's gravity cannot hold on to it. Although why this happens is understood, the details about how, and exactly where, the coronal gases are accelerated is one subject of ongoing investigation. The solar wind flows fastest when the sunspot cycle is at its minimum, because there is less turbulence in the corona to slow it down. Discussion of the solar wind's effects continues below.

Our Bubble of Interplanetary Space

The "vacuum" of interplanetary space includes copious amounts of energy radiated from the Sun, some interplanetary and interstellar dust (microscopic solid particles) and gas, and the solar wind. The solar wind, discovered by Eugene Parker in 1958, is a flow of lightweight ions and electrons (which together comprise plasma) thrown from the Sun.

The solar wind flows outward from our star at about 400 km per second (about 1 million miles per hour), measured in the vicinity of Earth's orbit. The Ulysses spacecraft found that it approximately doubles its speed at high solar latitudes. It has a visible effect on comet tails, blowing the charged-particle component of the comet's tail out away from the Sun as if they were wind socks. The solar wind inflates a bubble, called the heliosphere, in the surrounding interstellar medium (ISM).

But before it gets out to the heliopause, the solar wind slows to subsonic speeds, creating a termination shock. This appears at the perimeter of the circle in Figure 1.4. Its actual shape, whether roughly spherical or teardrop, depends on magnetic field strengths, as yet unknown. Humanity's most distant object the Voyager 1 spacecraft, still in routine communication with JPL, crossed the termination shock in December 2004 at a distance of 94 AU from the Sun. Today the craft is passing through the heliosheath, the region between the termination shock and the heliopause, at a rate of 3.6 AU per year, and is expected to reach the heliopause in about 2015 while still communicating with Earth.

In Figure 1.4, temperatures are theorized; none have been actually measured very far beyond the termination shock. Note that even with the high particle temperatures, their density is so low that massive objects like spacecraft do not "feel" these temperatures. Instead, their thermal states are dominated by solar (stellar) heating and exposure to shadow.

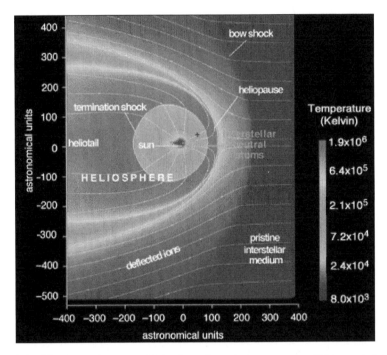

Figure 1.4: The boundary at which the solar wind meets the ISM, containing the collective "solar" wind from other local stars in our galaxy, is called the heliopause. This is where the solar wind and the Sun's magnetic field stop. The boundary is theorized to be roughly teardrop-shaped, because it gets "blown back" to form a heliotail, as the Sun moves through the ISM (toward the right in the diagram above). The Sun's relative motion may also create an advance bow shock, analogous to that of a moving boat. This is a matter of debate and depends partly on the strength of the interstellar magnetic field. Diagram courtesy Dr. Gary Zank, University of Delaware.

The white lines in Figure 1.4 represent charged particles, mostly hydrogen ions, in the interstellar wind. They are deflected around the heliosphere's edge (the heliopause). The pink arrow shows how neutral particles penetrate

View the video of Dr. Edward C. Stone's Caltech lecture, "The Voyager Journeys to Interstellar Space" (50 minutes).[12]

the heliopause. These are primarily hydrogen and helium atoms, which are mostly not affected by magnetic fields, and there are also heavier dust grains. These interstellar neutral particles make up a substantial fraction of the material found within the heliosphere. The little black + in the circular area represents the location where Voyager 1 was, at 80 AU from the Sun, in January 2001.

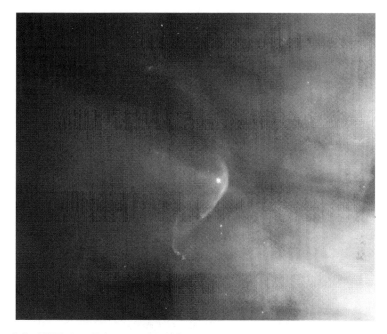

Figure 1.5: HST view February 1995. Arcing structure is stellar bow shock about half a light-year across, from star L.L. Orionis's wind colliding with the Orion Nebula flow. Courtesy NASA, Hubble Heritage Team (STScI/AURA).

The solar wind changes with the 11-year solar cycle, and the interstellar medium is not homogeneous, so the shape and size of the heliosphere probably fluctuate.

The solar magnetic field is the dominating magnetic field within the heliosphere, except in the immediate environment of those planets that have their own global magnetic fields. It can be measured by spacecraft through-out the solar system, but not here on Earth, where we are shielded by our planet's own magnetic field.

The actual properties of the interstellar medium (outside the helio-sphere), including the strength and orientation of its magnetic field, are important in determining the size and shape of the heliopause. Measure-ments that the two Voyager spacecraft will make in the region beyond the termination shock, and possibly beyond the heliopause, will provide impor-tant inputs to models of the termination shock and heliopause. Even though the Voyagers will sample these regions only in discrete locations, their in-formation will result in more robust overall models.

The Terrestrial Planets

The planets Mercury, Venus, Earth, and Mars, are called terrestrial because they have a compact, rocky surface like Earth's terra firma. The terrestrial planets are the four innermost planets in the solar system. None of the terrestrial planets have rings, although Earth does have belts of trapped radiation, as discussed below. Among the terrestrials, only Earth has a substantial planetary magnetic field. Mars and the Earth's Moon have lo-calized regional magnetic fields at different places across their surfaces, but no global field.

Of the terrestrial planets, Venus, Earth, and Mars have significant at-mospheres. The gases present in a planetary atmosphere are related to a planet's size, mass, temperature, how the planet was formed, and whether life is present. The temperature of gases may cause their molecules or atoms to achieve velocities that escape the planet's gravitational field. This con-tributes to Mercury's lack of a permanent atmosphere, as does its proximity to the source of the relentless solar wind.

The presence of life on Earth causes oxygen to be abundant in the at-mosphere, and in this Earth is unique in our solar system. Without life, most of the oxygen would soon become part of the compounds on the planet's surface. Thus, the discovery of oxygen's signature in the atmosphere of an extrasolar planet would be significant.

Mercury lacks an atmosphere to speak of. Even though most of its surface is very hot, there is strong evidence that water ice exists in loca-tions near its north and south poles which are kept permanently-shaded by crater walls. This evidence comes from Earth-based radar observations of the innermost planet. The discovery of permanently shaded ice at the poles of Earth's Moon strengthens arguments that the indications of ice on Mercury may be real.

Mercury was visited by Mariner 10 which flew by twice in 1974 and once in 1975, capturing images of one hemisphere. The Messenger spacecraft, which launched in 2004, entered into orbit around Mercury on 18 March 2011.

Venus's atmosphere of carbon dioxide is dense, hot, and permanently cloudy, making the planet's surface invisible. Its best surface studies have come from landers and imaging radar from orbiting spacecraft.

Venus has been visited by more than 20 spacecraft. The Magellan mission used synthetic aperture radar imaging and altimetry to map its surface at high resolution from 1990 to 1994. The European Venus Express, launched in 2005, has been orbiting Venus since April 2006.

Earth, as of March 2011, is still the only place known to harbor life. And life has flourished here since the planet was young. Our home planet is also unique in having large oceans of surface water, an oxygen-rich atmosphere, and shifting crustal sections floating on a hot mantle below, described by the theory of plate tectonics. Earth's Moon orbits the planet once every 27.3 days at an average distance of about 384,400 km. The Moon's orbital distance is steadily increasing at the very slow rate of 38 meters per millennium. Its distance at this point in its history makes the Moon appear in the sky to be about the same size as the Sun, subtending about half a degree.

Earth's Radiation Environment: JPL's first spacecraft, Explorer 1, carried a single scientific instrument, which was devised and operated by James Van Allen and his team from the University of Iowa. Early in 1958 the experiment discovered bands of rapidly moving charged particles trapped by Earth's magnetic field in toroidal (doughnut-shaped) regions surrounding the equator. The illustration below shows these belts only in two dimensions, as if they were sliced into thin cross-sections.

The belts that carry Van Allen's name have two areas of maximum density. The inner region, consisting largely of protons with an energy greater than 30 million eV, is centered about 3,000 km above Earth's surface. The outer belt is centered about 15,000 to 20,000 km up, and contains electrons with energies in the hundreds of millions of eV. It also has a high flux of

Figure 1.6: Van Allen belts in cross section, with the South Atlantic Anomaly.

protons, although of lower energies than those in the inner belt.

Flight within these belts can be dangerous to electronics and to humans because of the destructive effects the particles have as they penetrate microelectronic circuits or living cells. Most Earth-orbiting spacecraft are operated high enough, or low enough, to avoid the belts. The inner belt, however, has an annoying portion called the South Atlantic Anomaly (SAA) which extends down into low-earth-orbital altitudes. The SAA can be expected to cause problems with spacecraft that pass through it.

Mars's atmosphere, also carbon dioxide, is much thinner than Earth's, but it sustains wispy clouds of water vapor. Mars has polar caps of carbon dioxide ice and water ice. The planet's surface shows strong evidence for extensive water coverage in its distant past, as well as possible evidence for water flow in small springs during recent times.

Three dozen spacecraft have been targeted for Mars, although many failed to reach their destination. Two rovers, Spirit and Opportunity, are currently exploring the surface, while three craft currently operate in orbit: 2001 Mars Odyssey, Mars Express, and Mars Reconnaissance Orbiter.

Terrestrial Planet Data

Table 1.5 shows a comparison of terrestrial planet data. For complete current data on each planet, see the National Space Science Data Center, which is NASA's permanent archive for space science mission data:

    http://nssdc.gsfc.nasa.gov/planetary/factsheet

Table 1.5: Terrestrial Planet Data

	Mercury	Venus	Earth	Mars
Mean distance from Sun (AU)	0.387	0.723	1	1.524
Light minutes from Sun*	3.2	6.0	8.3	12.7
Mass (× Earth)	0.0553	0.815	1	0.107
Equatorial radius (× Earth) Rotation period	0.383	0.949	1	0.533
(Earth days)	175.942	−116.75 Retrograde	1	1.027
Orbit period (Earth years)	0.241	0.615	1	1.881
Mean orbital velocity (km/s)	47.87	35.02	29.78	24.13
Natural satellites	0	0	1	2
Surface atmospheric pressure (bars)	Near 0	92	1	.0069 to .009
Global Magnetic field	Faint	None	Yes	None

*Light minutes are useful for expressing distances within the solar system because they indicate the time required for radio communication with spacecraft at their distances.

Asteroids

Though there are several named groups of asteroids, which are covered in the next section, the term "asteroid" has increasingly come to particularly refer to the small rocky and metallic bodies of the inner solar system out to the orbit of Jupiter. Millions of "main-belt" asteroids orbit the Sun mostly between Mars and Jupiter. Asteroids are also called minor planets.

The Jovian Planets

Jupiter, Saturn, Uranus, and Neptune are known as the Jovian (Jupiter-like) planets, because they are all gigantic compared with Earth, and they have a gaseous nature like Jupiter's—mostly hydrogen, with some helium and trace gases and ices. The Jovian planets are thus referred to as the "gas giants" because gas is what they are mostly made of, although some or all of them probably have small solid cores. All have significant planetary magnetic fields, rings, and lots of satellites.

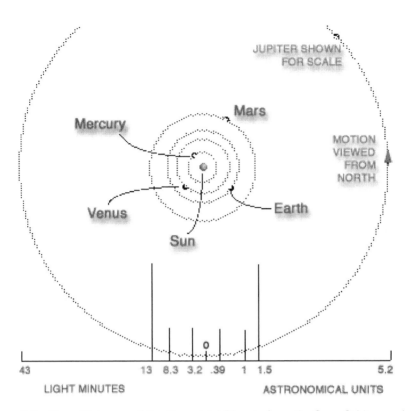

Figure 1.7: Mean Distances of the Terrestrial Planets from the Sun. Orbits are drawn approximately to scale.

Jupiter is more massive than all the other plan-
ets combined. It emits electromagnetic energy from
charged atomic particles spiraling through its strong
magnetic field. If this sizzling magnetosphere were
visible to our eyes, Jupiter would appear larger then
the full Moon in Earth's sky. The trapped radiation
belts near Jupiter present a hazard to spacecraft as do Earth's Van Allen
belts, although the Jovian particle flux and distribution differ from Earth's.
Bringing a spacecraft close to Jupiter presents a hazard mostly from ionized
particles. Spacecraft intended to fly close to Jupiter must be designed with
radiation-hardened components and shielding. Spacecraft using Jupiter for
gravity assist may also be exposed to a harsh radiation dose. Instruments

not intended to operate at Jupiter must be protected by being powered off or by having detectors covered.

One spacecraft, Galileo, has orbited Jupiter, and six others have flown by: Pioneer 10, Pioneer 11, Voyager 1, Voyager 2, Ulysses, and Cassini. New Horizons will fly by Jupiter on 28 February 2007 to absorb momentum from the planet and gain a boost to Pluto and beyond.

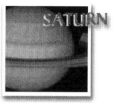

Saturn, the farthest planet easily visible to the unaided eye, is known for its extensive, complex system of rings, which are very impressive even in a small telescope. Using a small telescope one can also discern the planet's oblateness, or flattening at the poles. Continued study of Saturn's ring system can yield new understandings of orbital dynamics, applicable to any system of orbiting bodies, from newly forming solar systems to galaxies. Saturn's moons Titan, Enceladus, Iapetus, and others have proven to be extraordinarily interesting.

Pioneer 11 and the Voyagers flew by Saturn, and the Cassini spacecraft is currently studying the system from within Saturn orbit. The European Huygens Probe, carried by Cassini, executed a successful mission in Titan's atmosphere and on its surface 14 January 2005.

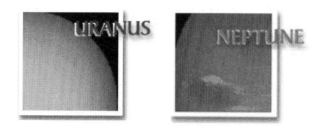

Uranus, which rotates on its side, and **Neptune** are of similar size and color, although Neptune seems to have a more active atmosphere despite its much greater distance from the Sun. Both planets are composed primarily of rock and various ices. Their extensive atmosphere, which makes up about 15% the mass of each planet, is hydrogen with a little helium. Both Uranus and Neptune have a retinue of diverse and interesting moons. These two cold and distant planets have had but one visitor, the intrepid Voyager 2.

Satellites of the Jovian Planets

The gas giants have numerous satellites, many of which are large, and seem as interesting as any planet. Small "new" satellites of the Jovian planets are being discovered every few years.

Figure 1.8: Jupiter's volcanic moon Io.

Jupiter's Galilean satellites, so named because Galileo Galilei discovered them in 1610, exhibit great diversity from each other. All four can be easily seen in a small telescope or binoculars. Io (Figure 1.8) is the closest of these to Jupiter. Io is the most volcanically active body in the solar system, due to heat resulting from tidal forces (discussed further in Chapter 3) which flex its crust. Powerful Earth-based telescopes can observe volcanoes resurfacing Io continuously.

Europa is covered with an extremely smooth shell of water ice. There is probably an ocean of liquid water below the shell, warmed by the same forces that heat Io's volcanoes.

Ganymede has mountains, valleys, craters, and cooled lava flows. Its ancient surface resembles Earth's Moon, and it is also suspected of having a sub-surface ocean.

Callisto, the outermost Galilean moon, is pocked all over with impact craters, indicating that its surface has changed little since the early days of its formation.

Figure 1.9: Saturn's planet-like moon Titan.

Saturn's largest moon, enigmatic Titan, is larger than the planet Mercury. Almost a terrestrial planet itself, Titan has a hazy nitrogen atmosphere denser than Earth's. The Huygens Probe's spectacularly successful mission in 2005 revealed rivers and lakebeds on the surface, and extensive details of its complex atmosphere.

Saturn also has many smaller and unexpectedly diverse satellites made largely of water ice. The "front," or leading, side of Saturn's icy satellite Iapetus is covered in dark material of some kind, and an equatorial mountain range as high as 13 kilometers was recently discovered on this 1450 km diameter moon.

Icy Enceladus orbits within the densest part of Saturn's E Ring, and has recently been shown to be the source of that ring's fine ice-particle makeup. Cassini spotted a system of crevasses near the south pole of Enceladus, and determined that the temperature in the crevasses is warmer than that of the rest of the 500-km diameter body.

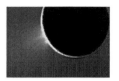

Figure 1.10: Geysers constantly erupt from Saturn's moon Enceladus.

In November 2005, backlit images showed fountains of water-ice particles spewing from Enceladus's south polar region. This activity is very surprising to observe on such a small icy satellite, and in fact remains unexplained as of early 2011.

All of Uranus's five largest moons have extremely different characteristics. The surface of Miranda, the smallest of these, shows evidence of extensive geologic activity. Umbriel's surface is dark, Titania and Ariel have trenches and faults, and Oberon's impact craters show bright rays similar to those on Jupiter's Callisto and our own Moon.

Neptune's largest moon Triton is partly covered with nitrogen ice and snow, and has currently active nitrogen geysers that leave sooty deposits on the surface downwind.

Rings

Jupiter's equatorial dust rings can be detected at close range in visible light and from Earth in the infrared. They show up best when viewed from behind, in forward scattered sunlight. Saturn, Uranus, and Neptune all have rings made up of myriad particles of ice ranging in size from dust and sand to boulders. Each particle in a ring is an individual satellite of the planet in its own right. Ring particles interact with each other in complex ways, affected by gravity and electrical charge. They also interact with the thin extended atmospheres of the planets.

Saturn's magnificent ring system, as visible in a small telescope from Earth, spans about 280,000 km, yet its thickness at any given point is less than 100 meters! The A-ring, measured at several points, was found to be only ten meters thick.

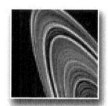

Figure 1.11: Saturn's rings in false color.[13]

There are some cases where two satellites occupy orbits very close to each other within a ring system, one satellite orbiting slightly farther from the planet than the ring, and the other satellite orbiting closer to the planet than the ring. The effect is that the satellites confine ring particles between their orbits into a narrow ring, by gravitationally interacting with the ring particles. These satellites are called shepherd moons. Examples are seen at Saturn and at Uranus.

Table 1.6: Jovian Planet Data (approximate)

	Jupiter	Saturn	Uranus	Neptune
Mean distance from Sun (AU)	5.20	9.58	19.20	30.05
Light hours from Sun*	0.72	1.3	2.7	4.2
Mass (× Earth)	317.8	95.2	14.5	17.1
Radius (× Earth)	11.21	9.45	4.01	3.88
Rotation period (hours)	9.9	10.7	17.2	16.1
Orbit period (Earth years)	11.9	29.4	83.7	163.7
Mean orbital velocity (km/s)	13.07	9.69	6.81	5.43
Known natural satellites (2007)	63	56	27	13
Rings	Dust	Extensive system	Thin, dark	Broken arcs

*Light time is useful for expressing distances within the solar system because it indicates the time required for radio communication with spacecraft at their distances.

Inferior and Superior Planets

Mercury and Venus are referred to as inferior planets, not because they are any less important, but because their orbits are closer to the Sun than is Earth's orbit. They always appear close to the Sun in Earth's morning or evening sky; their apparent angle from the Sun is called elongation. The outer planets, Mars, Jupiter, Saturn, Uranus, and Neptune, are all known as superior planets because their orbits are farther from the Sun than the Earth's.

Phases of Illumination

Inferior planets may pass between the Earth and the Sun on part of their orbits, so they can exhibit nearly the complete range of phases from the Earth's point of view, from the dark "new" phase, to slim "crescent" phase, to the mostly lit "gibbous" phase (approximating the fully illuminated "full" phase when approaching the other side of the Sun). Our own Moon, of course, exhibits all the phases.

Figure 1.13: An object in its "new" phase.

    Superior planets, though, usually appear gibbous, and appear full only when at opposition (see below), from our earthly point of view.

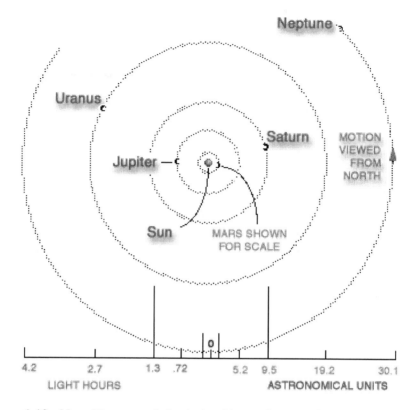

Figure 1.12: Mean Distances of the Jovian Planets from the Sun. Orbits are drawn approximately to scale.

Figure 1.13 shows a planet or a moon appearing as a crescent in its "new" phase because it is nearly between the observer and the Sun.

Viewed from superior planets, Earth goes through phases. Superior planets can be seen as crescents only from the vantage point of a spacecraft that is beyond them.

Conjunction, Transit, Occultation, Opposition

When two bodies appear to pass closest together in the sky, they are said to be in conjunction. When a planet passes closest to the Sun as seen from Earth and all three bodies are approximately in a straight line, the planet is said to be in solar conjunction. The inferior planets Venus and Mercury can have two kinds of conjunctions with the Sun: (1) An inferior conjunction, when the planet passes approximately between Earth and Sun (if it passes exactly between them, moving across the Sun's face as seen from Earth, it is said to be in transit); and (2) A superior conjunction when Earth and the other planet are on opposite sides of the Sun and all three bodies are again nearly in a straight line. If a planet disappears behind the Sun because the Sun is exactly between the planets, it is said to be in occultation.

Figure 1.14: Conjunctions, Elongations, and Opposition illustrated.

Superior planets can have only superior conjunctions with the Sun. At superior conjunction the outer planet appears near its completely illuminated full phase. The planet is said to be at opposition to the Sun when both it and the Earth are on the same side of the Sun, all three in line. (The Moon, when full, is in opposition to the Sun; the Earth is then approximately between them.)

Opposition is a good time to observe an outer planet with Earth-based instruments, because it is at its nearest point to the Earth and it is in its fullest phase.

Inferior planets can never be at opposition to the Sun, from Earth's point of view.

Occultations, transits, conjunctions, and oppositions offer special opportunities for scientific observations by spacecraft. Studies of the solar corona and tests of general relativity can be done at superior conjunctions. Superior conjunctions also present challenges communicating with a spacecraft nearly behind the Sun, which is overwhelmingly noisy at the same radio frequencies as those used for communications. At opposition, such radio noise is at a minimum, presenting ideal conditions for gravitational wave searches. These special opportunities and challenges are further discussed in later chapters.

The Minor Planets

Minor planets, also called asteroids, are rocky objects in orbit around the Sun. Most orbit in the main asteroid belt between Mars and Jupiter, moving in the same direction as the planets. They range in size from Ceres, which has a diameter of about 1000 km, down to the size of pebbles. Sixteen have a diameter of 240 km or greater.

Asteroids are classified according to their chemical composition, based on observed spectra and albedo (reflectivity). More than 75% are C-type asteroids which are dark and reddish with an albedo less than 0.10. They are similar to carbonaceous chondrite meteorites and exhibit about the same chemical composition as the Sun minus its volatiles. About 17% are S-type asteroids, which are brighter, with an albedo of 0.10 to 0.22. They are metallic nickel-iron mixed with iron- and magnesium-silicates. Most of the rest are M-type aster-

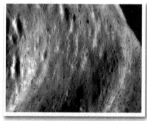

Figure 1.15: Asteroid 433 Eros imaged by NEAR-Shoemaker October 24, 2000 from orbit 100 km above surface.[14]

oids, with an albedo of 0.10 to 0.18, made of pure nickel-iron. There are several other rare-composition types of asteroids.

Eight spacecraft have crossed the main asteroid belt en route to their destinations, as of March 2011: Pioneers 10 and 11, Voyagers 1 and 2, Ulysses, Galileo (crossed twice), Cassini, and New Horizons. To their good fortune, none of them ever "discovered" any asteroids by colliding with them. Galileo made observations of two main-belt asteroids, 951 Gaspara and 243 Ida which was found to have its own satellite-asteroid. Cassini imaged the main-belt asteroid 2685 Masursky. (The number N before an asteroid's name means it was the Nth to have its orbit determined.)

There are relatively empty areas between the main concentrations of asteroids in the Main Belt called the Kirkwood gaps, where an object's orbital period would be a simple fraction of Jupiter's. Such a gravitational resonance, over time, causes objects to migrate into different orbits.

The "main belt" asteroids can actually be categorized as divided into two belts, according to confirming data from the current Sloan Digital Sky Survey. The inner belt, centered at 2.8 AU from the Sun, contains silicate-rich asteroids, and an outer belt, centered at 3.2 AU, contains asteroids rich in carbon.

The Main Asteroid Belt

Figure 1.16 illustrates the main belt of asteroids between the orbits of Mars and Jupiter. Some asteroids have orbits outside the main belt, either farther from or closer to the Sun than the main belt. Those which approach Earth are called Near Earth Asteroids, NEAs. Most of the objects which approach Earth are asteroids or "dead" comets, but a few are "live" comets. Together, these asteroids and comets are known as Near Earth Objects.

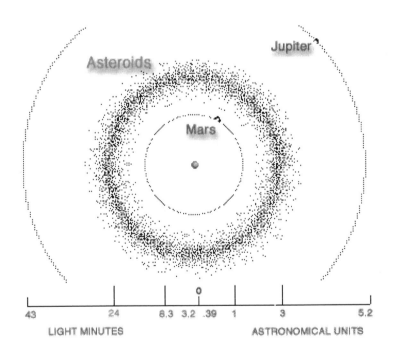

Figure 1.16: Main Belt Asteroids. Orbits are drawn approximately to scale.

Near Earth Objects (NEOs)

In terms of orbital elements, NEOs are asteroids and comets with perihelion distance q less than 1.3 AU. Near-Earth Comets (NECs) are further restricted to include only short-period comets (i.e., orbital period P less than 200 years). The vast majority of NEOs are asteroids (NEAs). NEAs are divided into groups Aten, Apollo, and Amor according to each one's perihelion distance (q), aphelion distance (Q) and semi-major axis (a). These terms are further discussed in Chapter 5.

Table 1.7: Classifications of Near-Earth Objects

NEO Group	Description	Definition
NECs	Near-Earth Comets	q < 1.3 AU, P< 200 years
NEAs	Near-Earth Asteroids	q < 1.3 AU
Atens	Earth-crossing NEAs with semi-major axes smaller than Earth's (named after 2062 Aten).	a < 1.0 AU, Q > 0.983 AU
Apollos	Earth-crossing NEAs with semi-major axes larger than Earth's (named after 1862 Apollo).	a > 1.0 AU, q < 1.017 AU
Amors	Earth-approaching NEAs with orbits exterior to Earth's but interior to Mars' (named after 1221 Amor).	a > 1.0 AU, q = 1.017 to 1.3 AU
PHAs	Potentially Hazardous Asteroids: NEAs whose Minimum Orbit Intersection Distances (MOID) with Earth are 0.05 AU or less and whose absolute magnitudes (H, a measure of brightness, and therefore size) are 22.0 or brighter.	MOID <= 0.05 AU, H <= 22.0

Other Asteroids

One asteroid has been discovered to date that orbits the Sun entirely within the Earth's orbit. Asteroid 2003 CP20 follows a highly inclined orbit in the area between Mercury and Earth. Its aphelion distance (furthest distance from the Sun), is 0.978 AU, less than the Earth's closest distance.

There are several hundred asteroids located near Jupiter's L4 and L5 Lagrange points (60 degrees ahead and 60 degrees behind Jupiter in its solar orbit), known as Trojans, named after heroes of the Trojan Wars. Mars has a Trojan, and there may be small Trojans in Lagrange points of Earth and Venus.

Small objects in the distant outer solar system might be similar to asteroids or comets; the definitions begin to be unclear. Asteroid-like objects between Saturn and Uranus are called Centaurs. Objects beyond Neptune are called trans-Neptunian objects, or Kuiper Belt objects.

Comets

Comets are formed of rocky material, dust, and water ice. Many have highly elliptical orbits that bring them very close to the Sun and swing them deep into space, often beyond the orbit of Pluto. Unlike the planets, which have orbits in nearly the same plane, comet orbits are oriented randomly in space.

The most widely accepted theory of the origin of comets is that there is a huge cloud of comets called the Oort Cloud (after the Dutch Astronomer Jan H. Oort who proposed the theory), of perhaps 10^{11} comets orbiting the Sun at a distance of about 50,000 AU (just under a light year). These comets are near the boundary between the gravitational forces of the Sun and the gravitational forces of other stars with which the Sun comes into interstellar proximity every several thousand years. According to the theory, these stellar passings perturb the orbits of the comets within the Oort cloud. As a result, some may be captured by the passing star, some may be lost to interstellar space, and some of their orbits are modified from a relatively circular orbit to an extremely elliptical one coming close to the Sun.

Another reservoir of comets is the Kuiper belt, a disk-shaped region about 30 to 100 AU from the Sun. This is considered to be the source of the short-period comets. The orbit of a Kuiper belt object is sometimes perturbed by gravitational interactions with the Jovian planets causing it to cross Neptune's orbit, where eventually it may have a close encounter with Neptune, either ejecting the comet or throwing it deeper into the solar system.

Comets have been known to break up on closest approach to the Sun. Discovered early in 1993, comet Shoemaker-Levy 9 had broken up apparently because of a close passage to Jupiter. It had been captured into orbit about Jupiter and eventually collided with the planet in July of 1994. The spectacular collision was widely observed. The SOHO spacecraft sometimes watches comets pass by the Sun. A time-lapse movie shows two comets actually colliding with our central star.[15]

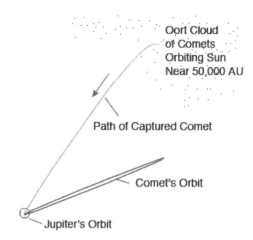

Figure 1.17: Capture and Orbit of a Typical Comet.

Figure 1.17 depicts a vast system, not drawn to scale. The Sun is at the center of the small circle that represents the orbit of Jupiter at 5 AU. The figure is not drawn to scale; the Oort Cloud would be about 10,000 times farther out than depicted here.

Comets are practically invisible until they come near the Sun and develop an extended structure. These structures are diverse and very dynamic, but they all include a surrounding cloud of diffuse material, called a coma, that usually grows in size and brightness as the comet approaches the Sun. The dense, inner coma often appears pointlike, but the central body, called the nucleus is rarely seen from Earth because it is too small and dim. The coma and the nucleus together constitute the head of the comet.

As some comets approach the Sun they develop enormous tails of luminous material that extend for millions of kilometers from the head, away from the Sun. When far from the Sun, the nucleus is very cold and its material is frozen solid. In this state comets are sometimes referred to as "dirty icebergs" or "dirty snowballs," since over half of their material is ice. Approaching within a few AU of the Sun, the surface of the nucleus begins to warm, and the volatiles evaporate. The evaporated molecules boil off and carry small solid particles with them, forming the comet's coma of gas and dust.

When a coma develops, dust reflects sunlight, while gas in the coma absorbs ultraviolet radiation and begins to fluoresce. At about 5 AU from the Sun, fluorescence usually becomes more intense than the reflected light.

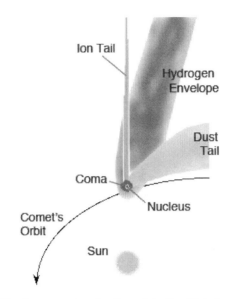

Figure 1.18: Components of a Comet in the Vicinity of the Sun.

As the comet absorbs ultraviolet light, chemical processes release hydrogen, which escapes the comet's gravity and forms a hydrogen envelope. This envelope cannot be seen from Earth because its light is absorbed by our atmosphere, but it has been detected by spacecraft.

The Sun's radiation pressure and solar wind accelerate materials away from the comet's head at differing velocities according to the size and mass of the materials. Thus, relatively massive dust tails are accelerated slowly and tend to be curved. The ion tail is much less massive, and is accelerated so greatly that it appears as a nearly straight line extending away from the comet opposite the Sun. This is clearly visible in the time-lapse movies available on the SOHO website.

Each time a comet visits the Sun, it loses some of its volatiles. Eventually, it becomes just another rocky mass in the solar system. For this reason, comets are said to be short-lived, on a cosmological time scale. Many believe that some asteroids are extinct comet nuclei, comets that have lost all of their volatiles.

Meteoroids, Meteors, Meteorites

Meteoroids are small, often microscopic, solid particles orbiting the Sun. We see them as meteors ("shooting stars" or "falling stars") when they enter Earth's atmosphere at tens of kilometers per second and burn up. Heat is generated by compression as the object plows into the atmosphere much faster than the speed of sound, causing the air near the object to glow, while some heat is transferred to the moving object. On almost any dark night, at least a few meteors may be seen. There are many more during several yearly meteor showers. Some display impressive fireballs, leaving cloudy trails behind. Any part of a meteor that reaches the ground is called a meteorite.

As volatiles boil or sublimate from comets, they carry small solid particles with them. Particles released from comets in this way becomes a source for meteoroids, causing meteor showers as the Earth passes through them. Since Earth and meteoroid orbits have a fixed relationship, meteor showers appear to originate at a point in the sky, as seen from Earth's surface, called the "radiant." For example, each autumn observers see the Leonoid meteors appear to radiate from within the constellation Leo. Meteoroids also come from the asteroid belt. Some rare meteoroids are actually debris lofted from the Moon or Mars as the result of large impacts on those bodies.

Figure 1.19: Meteors in the 1966 Leonid shower.

Notes

[1] http://www.haystack.mit.edu/edu/undergrad/materials/tut4.html

[2] http://www.nas.edu/history/igy

[3] Look for "Resolution GA26-5-6" at http://www.iau.org/static/resolutions.

[4] http://planetquest.jpl.nasa.gov

[5] Image ID: STScI-PR94-01

[6] http://soho.nascom.nasa.gov

[7] http://stereo.gsfc.nasa.gov

[8] http://zebu.uoregon.edu/~soper/Sun/mass.html

[9] http://stars.astro.illinois.edu/sow/sun.html

[10] http://nssdc.gsfc.nasa.gov/planetary/factsheet/sunfact.html

[11] http://www.jpl.nasa.gov/basics/C3Aug99.mov

[12] http://today.caltech.edu/theater/328_56k.ram

[13]Voyager-2's highly enhanced color view, obtained August 17, 1981. Image PIA01486 courtesy JPL/NASA.

[14]Image courtesy Johns Hopkins University Applied Physics Laboratory.

[15]http://sohowww.nascom.nasa.gov/bestofsoho

Chapter 2

Reference Systems

> **Objectives:** Upon completion of this chapter you will be able to describe the system of terrestrial coördinates, the rotation of Earth, precession, nutation, and the revolution of Earth about the Sun. You will be able to describe how the locations of celestial objects are stated in the coördinate systems of the celestial sphere. You will be able to describe the use of epochs and various conventions of timekeeping.

Spatial coördinates and timing conventions are adopted in order to consistently identify locations and motions of an observer, of natural objects in the solar system, and of spacecraft traversing interplanetary space or orbiting planets or other bodies. Without these conventions it would be impossible to navigate the solar system.

Terrestrial Coördinates

A great circle is an imaginary circle on the surface of a sphere whose center is the center of the sphere. Great circles that pass through both the north and south poles are called meridians, or lines of longitude. For any point on the surface of Earth a meridian can be defined.

The prime meridian, the starting point measuring the east-west locations of other meridians, marks the site of the old Royal Observatory in Greenwich, England. Longitude is expressed in degrees, minutes, and seconds of arc from 0 to 180 degrees eastward or westward from the prime meridian. For example,

Figure 2.1: Lines of longitude (above) and latitude (below).

45

downtown Pasadena, California, is located at 118 degrees, 8 minutes, 41 seconds of arc west of the prime meridian: 118 8' 41" W.

The starting point for measuring north-south locations on Earth is the equator, a great circle which is everywhere equidistant from the poles. Circles in planes parallel to the equator define north-south measurements called parallels, or lines of latitude. Latitude is expressed as an arc subtended between the equator and the parallel, as seen from the center of the Earth. Downtown Pasadena is located at 34 degrees, 08 minutes, 44 seconds latitude north of the equator: 34 08' 44" N.

Throughout the history of navigation, determining one's latitude on the Earth's surface has been relatively easy. In the northern hemisphere for example, simply measuring the height of the star Polaris above the horizon results in a fairly close approximation of one's latitude. Measurement of longitude, however, has been a historically siginificant endeavor, since its determination requires accurate timekeeping. John Harrison (1693-1776) eventually succeeded in developing a chronometer good enough to do the trick.[1]

One degree of latitude equals approximately 111 km on the Earth's surface, and by definition exactly 60 nautical miles. Because meridians converge at the poles, the length of a degree of longitude varies from 111 km at the equator to 0 km at the poles where longitude becomes a point.

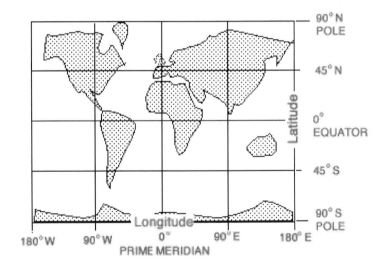

Figure 2.2: Terrestrial Coördinates Grid.

Rotation and Revolution

"Rotation" refers to an object's spinning motion about its own axis. "Revolution" refers the object's orbital motion around another object. For example, Earth rotates on its own axis, producing the 24-hour day. Earth revolves about the Sun, producing the 365-day year. A satellite revolves around a planet.

Earth's Rotation

The Earth rotates on its axis relative to the Sun every 24.0 hours mean solar time, with an inclination of 23.45 degrees from the plane of its orbit around the Sun. Mean solar time represents an average of the variations caused by Earth's non-circular orbit. Its rotation relative to "fixed" stars (sidereal time) is 3 minutes 56.55 seconds shorter than the mean solar day, the equivalent of one solar day per year.

Precession of Earth's Axis

Forces associated with the rotation of Earth cause the planet to be slightly oblate, displaying a bulge at the equator. The moon's gravity primarily, and to a lesser degree the Sun's gravity, act on Earth's oblateness to move the axis perpendicular to the plane of Earth's orbit. However, due to gyroscopic action, Earth's poles do not "right themselves" to a position perpendicular to the orbital plane. Instead, they precess at 90 degrees to the force applied. This precession causes the axis of Earth to describe a circle having a 23.4 degree radius relative to a fixed point in space over about 26,000 years, a slow wobble reminiscent of the axis of a spinning top swinging around before it falls over.

Because of the precession of the poles over 26,000 years, all the stars, and other celestial objects, appear to shift west to east at the rate of .014 degree each year (360 degrees in 26,000 years). This apparent motion is the main reason for astronomers as well as spacecraft operators to refer to a common epoch such as J2000.0.

At the present time in Earth's 26,000 year precession cycle, a bright star happens to be very close, less than a degree, from the north celestial pole. This star is called Polaris, or the North Star.

Stars do have their own real motion, called proper motion. In our vicinity of the galaxy, only a few bright stars exhibit a large enough proper

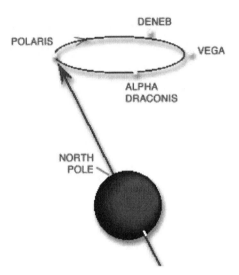

Figure 2.3: Precession of Earth's Axis Over 26,000 Years.

motion to measure over the course of a human lifetime, so their motion does not generally enter into spacecraft navigation. Because of their immense distance, stars can be treated as though they are references fixed in space. Some stars at the center of our galaxy, though, display tremendous proper motion speeds as they orbit close to the massive black hole located there.[2]

Nutation

Superimposed on the 26,000-year precession is a small nodding motion with a period of 18.6 years and an amplitude of 9.2 arc seconds. This nutation can trace its cause to the 5 degree difference between the plane of the Moon's orbit, the plane of the Earth's orbit, and the gravitational tug on one other.

Revolution of Earth

Earth revolves in orbit around the Sun in 365 days, 6 hours, 9 minutes with reference to the stars, at a speed ranging from 29.29 to 30.29 km/s. The 6 hours, 9 minutes adds up to about an extra day every fourth year, which is designated a leap year, with the extra day added as February 29th. Earth's orbit is elliptical and reaches its closest approach to the Sun, a perihelion of 147,090,000 km, on about January fourth of each year. Aphelion comes six months later at 152,100,000 km.

Shorter-term Polar Motion

Aside from the long-term motions, the Earth's rotational axis and poles have two shorter periodic motions. One, called the Chandler wobble, is a free nutation with a period of about 435 days. There is also a yearly circular motion, and a steady drift toward the west caused by fluid motions in the Earth's mantle and on the surface. These motions are tracked by the International Earth Rotation and Reference Systems Service, IERS. The IERS Earth Orientation Center's latest displays are online.[3]

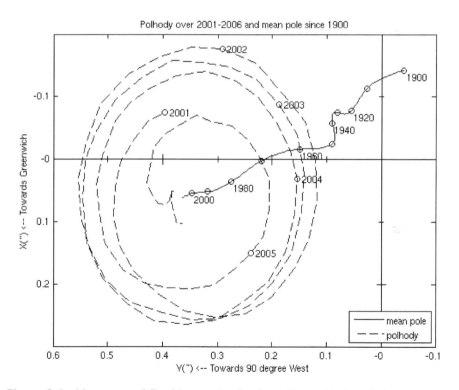

Figure 2.4: Movement of Earth's rotational poles 2001 to 2006, and the mean pole location from the year 1900 to 2000. Units are milliarcseconds. Image courtesy IERS Earth Orientation Center.

Epochs

Because we make observations from Earth, knowledge of Earth's natural motions is essential. As described above, our planet rotates on its axis daily and revolves around the Sun annually. Its axis precesses and nutates. Even the "fixed" stars move about on their own. Considering all these motions, a useful coördinate system for locating stars, planets, and spacecraft must be pinned to a single snapshot in time. This snapshot is called an epoch. By convention, the epoch in use today is called J2000.0, which refers to the mean equator and equinox of year 2000, nominally January 1st 12:00 hours Universal Time (UT). The "J" means Julian year, which is 365.25 days long. Only the 26,000-year precession part of the whole precession/nutation effect is considered, defining the mean equator and equinox for the epoch.

The last epoch in use previously was B1950.0 - the mean equator and equinox of 1949 December 31st 22:09 UT, the "B" meaning Besselian year, the fictitious solar year introduced by F. W. Bessell in the nineteenth century. Equations are published for interpreting data based on past and present epochs.

Making Sense

Given an understanding of the Earth's suite of motions – rotation on axis, precession, nutation, short-term polar motions, and revolution around the Sun – and given knowledge of an observer's location in latitude and longitude, meaningful observations can be made. For example, to measure the precise speed of a spacecraft flying to Saturn, you have to know exactly where you are on the Earth's surface as you make the measurement, and then subtract out the Earth's motions from that measurement to obtain the spacecraft's speed.

The Celestial Sphere

A useful construct for describing locations of objects in the sky is the celestial sphere, which is considered to have an infinite radius. The center of the Earth is the center of the celestial sphere, and the sphere's pole and equatorial plane are coincident with those of the Earth. See Figure 2.5. We can specify precise location of objects on the celestial sphere by giving the celestial equivalent of their latitudes and longitudes.

The point on the celestial sphere directly overhead for an observer is the zenith. An imaginary arc passing through the celestial poles and through the zenith is called the observer's meridian. The nadir is the direction opposite the zenith: for example, straight down from a spacecraft to the center of the planet.

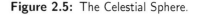

Figure 2.5: The Celestial Sphere.

Declination, Right Ascension

Declination (DEC) is the celestial sphere's equivalent of latitude and it is expressed in degrees, as is latitude. For DEC, + and − refer to north and south, respectively. The celestial equator is 0 DEC, and the poles are +90 and –90.

Right ascension (RA) is the celestial equivalent of longitude. RA can be expressed in degrees, but it is more common to specify it in hours, minutes, and seconds of time: the sky appears to turn 360 in 24 hours, or 15 in one hour. So an hour of RA equals 15 of sky rotation.

Another important feature intersecting the celestial sphere is the ecliptic plane. This is the plane in which the Earth orbits the Sun, 23.4 from the celestial equator. The great circle marking the intersection of the ecliptic plane on the celestial sphere is where the Sun and planets appear to travel, and it's where the Sun and Moon converge during their eclipses (hence the name).

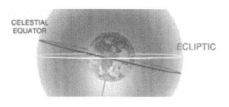

Figure 2.6: The Ecliptic Plane.

The zero point for RA is one of the points where the ecliptic circle intersects the celestial equator circle. It's defined to be the point where the Sun crosses into the northern hemisphere beginning spring: the vernal equinox. Also known as the first point of Aries, it is often identified by the symbol of the ram: ♈

The equinoxes are times at which the center of the Sun is directly above the equator, marking the beginning of spring and autumn. The day and night would be of equal length at that time, if the Sun were a point and not a disc, and if there were no atmospheric refraction. Given the apparent disc of the Sun, and the refraction, day and night actually become equal at a point within a few days of each equinox. The RA and DEC of an object specify its position uniquely on the celestial sphere just as the latitude and longitude do for an object on the Earth's surface. For example, the very bright star Sirius has celestial coördinates 6 hr 45 min RA and -16 43' DEC.

The International Celestial Reference System

The International Celestial Reference System (ICRS) is the fundamental celestial reference system that has been adopted by the International Astronomical Union (IAU) for high-precision positional astronomy. The ICRS, with its origin at the solar system barycenter and "space fixed" axis directions, is meant to represent the most appropriate coördinate system for positions and motions of celestial objects. RA and DEC measurements can be transformed to the ICRS sytem, which is compatible with the J2000.0-based system. The reference frame created by the ICRS is called the International Celestial Reference Frame, ICRF.[4]

HA-DEC versus AZ-EL Radio Telescopes

The discussion gets a little more involved here, but this section serves only to explain the old design for Deep Space Network antennas, as well as large optical and radio telescopes, and why it all changed not too long ago.

Before you can use RA and DEC to point to an object in the sky, you have to know where the RA is at present for your location, since the Earth's rotation continuously moves the fixed stars (and their RA) with respect to your horizon. If the RA of the object happens to place it overhead on your meridian, you're fine. But it probably isn't, so you determine the object's hour angle (HA), which is the distance in hours, minutes, and seconds westward along the celestial equator from the observer's meridian to the object's RA. In effect, HA represents the RA for a particular location and time of day. HA is zero when the object is on your meridian.

Older radio telescopes were designed with one mechanical axis parallel to Earth's axis. To track an interplanetary spacecraft, the telescope would point to the spacecraft's known HA and DEC, and then for the rest of the tracking period it would simply rotate in HA about the tilted axis (called its polar axis), as the Earth turns. This kind of mounting is traditionally called an equatorial mount when used for optical telescopes. It's a fine mount for a small instrument, but unsuited to very heavy structures because the tilted polar bearing has to sustain large asymmetric loads.

These loads include not only the whole reflector dish, but also an HA counterweight heavy enough to balance the antenna, the DEC bearing, and its DEC counterweight! Also the structure has to be designed specifically for its location, since the polar bearing's angle depends on the station's latitude. This image shows the first Deep Space Network (DSN) antenna installed at the Canberra, Australia site, looking down along the polar bearing, which is the axis of the antenna's large central wheel. This HA-DEC antenna is no longer in service, nor is its sister at the DSN site at Madrid, Spain. Its counterpart at the Goldstone, California site has been converted to a radio telescope dedicated to educational use.[5] An enlarged and annotated view of its complex design can be seen online.[6]

Figure 2.7: An older DSN station with HA-DEC mount (DSS-42).

A simpler system was needed for larger Deep Space Network antennas. The solution is an azimuth-elevation configuration. The design permits mechanical loads to be symmetric, resulting in less cumbersome, less expensive hardware that is easier to maintain.

It locates a point in the sky by elevation (EL) in degrees above the horizon, and azimuth (AZ) in degrees clockwise (eastward) from true north. These coördinates are derived from published RA and DEC by computer programs. This computerization was the key that permitted the complex mechanical structures to be simplified.

In an AZ-EL system anywhere on Earth, east is 90 degrees AZ, and halfway up in EL or altitude (ALT) would be 45 degrees. AZ-EL and ALT-AZ are simply different names for the same reference system, ALTitude being the same measurement as ELevation.

Figure 2.8: DSN station with AZ-EL mount (DSS-14).

Figure 2.8 shows a DSN antenna at Goldstone that has a 70-meter aperture, over twice that of the Australian HA-DEC antenna shown above. In the picture it is pointing to an EL around 10. The EL bearing is located at the apex of the triangular support visible near the middle right of the image. The whole structure rotates in AZ clockwise or counterclockwise atop the large cylindrical concrete pedestal. It is pointing generally east in the image (around 90°azimuth), probably beginning to track a distant spacecraft as it rises over the desert horizon. All newly designed radio telescopes use the AZ-EL system.

Then There's X-Y

To complete our survey of mounting schemes for DSN antennas (including steerable non-DSN radio telescopes as well), we need to describe the X-Y mount. Like AZ-EL, the X-Y mount also has two perpendicular axes. By examining the image of DSS16 here you can see that it cannot, however, directly swivel in azimuth as can the AZ-EL (ALT-AZ) -mounted antenna. But the X-Y mount has advantages over AZ-EL.

Its first advantage is that it can rotate freely in any direction from its upward-looking zenith central position without any cable wrap-up issues anywhere within its view.

The other advantage is a matter of keyholes. A "keyhole" is an area in the sky where an antenna cannot track a spacecraft because the required angular rates would be too high. Mechanical limitations may also contribute to keyhole size, for example the 70-m antennas are not allowed to track above 88 elevation. For an HA-DEC antenna, the keyhole is large and, in the northern hemisphere, is centered near the North Star. To track through that area the antenna would have to whip around prohibitively fast in hour angle.

Figure 2.9: DSN station with X-Y mount (DSS-16).

Imagine an AZ-EL antenna like the 70-m DSS in the section above. If a spacecraft were to pass directly overhead, the AZ-EL antenna would rise in elevation until it reached its straight-up maximum near 90. But then the antenna would have to whip around rapidly in azimuth as the spacecraft is first on the east side of the antenna, and then a moment later is on the west. The antennas' slew rate isn't fast enough to track that way, so there would be an interruption in tracking until acquiring on the other side. (AZ-EL antennas in the DSN aren't designed to bend over backwards, or "plunge" in elevation.)

The X-Y antenna is mechanically similar to the old HA-DEC antenna, but with its "polar" axis laid down horizontally, and not necessarily aligned to a cardinal direction. The X-Y antenna is situated so that its keyholes (two of them) are at the eastern and western horizon. This leaves the whole sky open for tracking spacecraft without needing impossibly high angular rates around either axis... it can bend over backwards and every which way. The X-Y's were first built for tracking Earth-orbiting spacecraft that require high angular rates and overhead passes. Earth-orbiters usually have an inclination that avoids the east and west keyholes, as well. Interplanetary spacecraft typically do not pass overhead, but rather stay near the ecliptic plane in most cases. Of course X-Y's can be used with interplanetary spacecraft also, but in the DSN they are only equipped with a 26-m aperture, smaller than most other DSN stations, and thus not useful for most interplanetary craft.

Time Conventions

Various expressions of time are commonly used in interplanetary space flight operations:

- UTC, Coördinated Universal Time, is the world-wide scientific standard of timekeeping. It is based upon carefully maintained atomic clocks and is highly stable. Its rate does not change by more than about 100 picoseconds per day. The addition or subtraction of leap seconds, as necessary, at two opportunities every year adjusts UTC for irregularities in Earth's rotation. UTC is used by astronomers, navigators, the Deep Space Network (DSN), and other scientific disciplines. Its reference point is Greenwich, England: when it is midnight there on Earth's prime meridian, it is midnight (00:00:00.000000) – "all balls" – UTC. The U.S. Naval Observatory website provides information in depth on the derivation of UTC.[7]

- UT, Universal Time also called Zulu (Z) time, was previously called Greenwich Mean Time, GMT. It is based on the imaginary "mean sun," which averages out the effects on the length of the solar day caused by Earth's slightly non-circular orbit about the Sun. UT is not updated with leap seconds as is UTC. Its reference point is also Greenwich, England: when it is noon on the prime meridian, it is noon (12:00:00) UT. It is common to see outdated references to GMT, even in currently operating flight projects. It is also common to encounter references to UT or GMT when the system actually in use is UTC, for example, "Uplink the command at 1801Z."

- Local time is UT adjusted for location around the Earth in time zones. Its reference point is one's immediate locality: when it is 12:00:00 noon Pacific Time at JPL, it is 20:00:00 UTC, and 13:00:00 Mountain Time in Denver, Colorado. Many locations change between standard time and daylight saving time (see below). Local time is also determined on other planets when needed.

  Local time on another planet is conceived as the equivalent value of time for the Sun's distance from the meridian, as it is on Earth. A planet that rotates more slowly than Earth would have an object in its sky at 1:00 local time move to 2:00 local time in more than an hour of Earth-clock time. Around 11:30 am or 12:30 pm at a particular location on Venus, the Sun would be nearly overhead. At 5:00 pm at a particular location on Mars, the Sun would be low in the west.

- TRM, Transmission time is the UTC time of uplink from Earth.

- OWLT, One-Way Light Time is the elapsed time it takes for light, or a radio signal, to reach a spacecraft or other body from Earth (or vice versa). Knowledge of OWLT is maintained to an accuracy of milliseconds. OWLT varies continuously as the spacecraft's distance from the Earth changes. Its reference points are the center of the Earth and the immediate position of a spacecraft or the center of a celestial body.

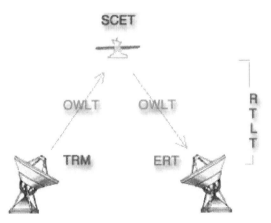

Figure 2.10: TRM + OWLT = SCET; TRM + RTLT = ERT.

- SCET, Spacecraft Event Time is the UTC time onboard the spacecraft. It is equal to TRM + OWLT. ERT is equal to SCET + OWLT.

- SCLK, Spacecraft Clock is the value of a counter onboard a spacecraft, described further in Chapter 11. SCLK has a nearly-direct relationship with SCET: it is the best possible on-board estimate of SCET. SCLK is not as constant and stable as the UTC-derived SCET. Its units of measurement are different from SCET.

  Tracking and predicting the exact relationship between SCLK and SCET is accomplished by analyzing telemetered SCLK values and trends with respect to the UTC-derived SCET, and regularly producing and applying a SCLK/SCET coefficients file which tracks the gradual drift of SCLK versus SCET.

- RTLT, Round-Trip Light Time is the elapsed time it takes for a signal to travel from Earth, be received and immediately transmitted or reflected by a spacecraft or other body, and return to the starting point. It is roughly equal to $2 \times$ OWLT, but not exactly, because of the different amount of distance the signal must travel on each leg due to the constant motions of both Earth and spacecraft. For reference, RTLT from here to the Moon is around 3 seconds, to the Sun, about 17 minutes. Voyager 1's RTLT at this writing in October 2000 is over 22 hours and increasing roughly an hour per year.

- ERT, Earth-Received Time is the UTC of an event received at a DSN station. One more definition may be useful as background information:

- DT, Dynamical Time, has replaced Ephemeris Time, ET, as the independent argument in dynamical theories and ephemerides. Its unit of duration is based on the orbital motions of the Earth, Moon, and planets. DT has two expressions, Terrestrial Time, TT, (or Terrestrial Dynamical Time, TDT), and Barycentric Dynamical Time, TDB. More information on these, and still more timekeeping expressions, may be found at the U.S. Naval Observatory website.[8]

  It is common to see outdated references to ET when a DT expression is intended, even in currently operating flight projects.

Figure 2.11 is an excerpt from a flight project's Sequence of Events (SOE). It illustrates use of UTC, ERT, TRM, OWLT, RTLT, SCET, and SCLK. (The SOE in general is discussed further in Chapter 15.)

The first vertical column in this SOE is a line item number. The next column specifies the item's UTC ground time, and the next column indicates whether that time is ERT or TRM. All on this page are ERT. Any items involving the DSN transmitter would appear as TRM. The COMMAND column specifies the command being executed on the spacecraft from the command sequence stored in the spacecraft's memory. The last column on the right shows the SCET, followed by the corresponding SCLK value, at which the command executes.

Daylight Saving Time

Daylight Saving Time begins for most of the United States at 2 am local on the second Sunday in March (as of 2007). Standard time returns at 2 am on the first Sunday in November (note that different time zones are switching at different moments). Hawaii, and Arizona (with the exception of the Navajo Nation) and the territories of Puerto Rico, Virgin Islands, Guam, and American Samoa do not observe DST but remain on Standard Time.

For the European Union, Summer Time, the equivalent of the U.S. Daylight Saving Time, begins at 0100 UTC on the last Sunday in March, and ends at 0100 UTC on the last Sunday in October (all time zones switch at the same moment in the EU).

Notes

[1] http://www.nmm.ac.uk/harrison
[2] http://www.mpe.mpg.de/ir/GC
[3] http://hpiers.obspm.fr/eop-pc
[4] http://hpiers.obspm.fr/webiers/icrf/icrf.html
[5] http://www.lewiscenter.org/gavrt
[6] www.jpl.nasa.gov/basics/34std.gif
[7] http://tycho.usno.navy.mil
[8] http://tycho.usno.navy.mil/systime.html

U.S. Government sponsorship under NASA Contract NAS7-1270 is acknowledged.

```
**********      SEQUENCE OF EVENTS:    YEAR-DAY OF YEAR  --> 2000-331                                 PAGE 321
** CASSINI **   S/C = 082              INPUT  FILE NAME  --> b023Od.pef
**********      SEQ = b023Od           OUTPUT FILE NAME  --> b023Od.soe
```

ITEM NO	UTC GND TIME DOY HH:MM:SS	T B	ACTION	EVENT DESCRIPTION	DSN	COMMAND (%=DCMD)	S/C EVENT TIME S/C CLOCK
4237	331 03:44:56	E		DEFINE ROTATIONAL DELTA OFFSET IN BASE ATTITUDE COORDINATES X: 0.0 MRAD Y: 0.0 MRAD Z: -11.3 MRAD		7DELTA_BASE	331 03:13:04 1353900257:073
4238	331 04:04:12	E		TURN OFF CDA ARTICULATION MECHANISM STEPPER MOTOR ELECTRONICS		79AM_MOTOR_PWR	331 03:32:20 1353901413:073
4239	331 04:04:22	E		PLACE COSMIC DUST ANALYZER IN SLEEP MODE OPERATIONS		79RT_SLEEP	331 03:32:30 1353901423:073
4240	331 04:04:32	E		INITIATE COSMIC DUST ANALYZER MEASUREMENT CYCLE USING ALL ON CHANNELS		79EVENT_DEFINE	331 03:32:40 1353901433:073
4241	331 04:05:32	E		INITIATE TEST PULSE AND AMPLITUDE LEVEL FOR COSMIC DUST ANALYZER TEST PULSE TYPE = 2 MP AMPLITUDE = 0		79DA_TEST_PULSE	331 03:33:40 1353901493:073
4242	331 04:06:32	E		INITIATE TEST PULSE AND AMPLITUDE LEVEL FOR COSMIC DUST ANALYZER TEST PULSE TYPE = 3 MP AMPLITUDE = 0		79DA_TEST_PULSE	331 03:34:40 1353901553:073
4243	331 04:47:54	E		SET SPACECRAFT OFFSET TURN RATE AND ACCELERATION PARAMETERS: TURN RATE X: 1.8 Y: 1.9 Z: 3.3 MRAD/S ACCEL X: 0.01 Y: 0.011 Z: 0.019 MRAD/S2		7PROFILE	331 04:16:02 1353904035:077
4244	331 04:47:56	E		DEFINE ROTATIONAL ABSOLUTE OFFSET IN BASE ATTITUDE COORDINATES X: 2.15 MRAD Y: 0.0 MRAD Z: 2.15 MRAD		7OFFSET	331 04:16:04 1353904037:077
4245	331 04:50:56	E		CHANGE SPACECRAFT TELEMETRY MODE TO S_N_ER_3		6CHG_SC_TM_IMM	331 04:19:04 1353904217:077
4246	331 04:50:56	E		EXECUTE BOTH CAMERAS COMMAND ID: 440		36NAC_TRIGGER	331 04:19:04 1353904217:078
4247	331 04:50:56	E		EXECUTE BOTH CAMERAS COMMAND ID: 440		36WAC_TRIGGER	331 04:19:04 1353904217:079
4248	331 04:50:56	E		START EXECUTION OF CIRS COMMAND SEQUENCE TABLE 91 RTI EXECUTE : 0 RTT CONTROL : IMMEDIATE ATT CONTROL : RELATIVE LOOP COUNT : 1 REL EXEC TIME: 0 MSEC ABS EXEC TIME: 2001-001T00:00:00.000		89EXE_CMD_SEQ	331 04:19:04 1353904217:080
4249	331 04:50:57	E		SET SPACECRAFT OFFSET TURN RATE AND ACCELERATION PARAMETERS: TURN RATE X: 1.8 Y: 1.9 Z: 3.3 MRAD/S ACCEL X: 0.01 Y: 0.011 Z: 0.019 MRAD/S2		7PROFILE	331 04:19:05 1353904218:079
4250	331 04:56:43	E		DEFINE ROTATIONAL DELTA OFFSET IN BASE ATTITUDE COORDINATES X: 0.0 MRAD Y: 0.0 MRAD Z: -4.3 MRAD		7DELTA_BASE	331 04:24:51 1353904564:080

Figure 2.11: Excerpt from a flight project's Sequence of Events (SOE).

Chapter 3

Gravitation and Mechanics

Objectives: Upon completion of this chapter you will be able to describe the force of gravity, characteristics of ellipses, and the concepts of Newton's principles of mechanics. You will be able to recognize acceleration in orbit and explain Kepler's laws in general terms. You will be able to describe tidal effect and how it is important in planetary systems.

Gravitation is the mutual attraction of all masses in the universe. While its effect decreases in proportion to distance squared, it nonetheless applies, to some extent, regardless of the sizes of the masses or their distance apart.

The concepts associated with planetary motions developed by Johannes Kepler (1571-1630) describe the positions and motions of objects in our solar system. Isaac Newton (1643-1727) later explained why Kepler's laws worked, by showing they depend on gravitation. Albert Einstein (1879-1955) posed an explanation of how gravitation works in his general theory of relativity.[1]

In any solar system, planetary motions are orbits gravitationally bound to a star. Since orbits are ellipses, a review of ellipses follows.

Ellipses

An ellipse is a closed plane curve generated in such a way that the sum of its distances from two fixed points (called the foci) is constant. In the illustration below, the sum of Distance A + Distance B is constant for any point on the curve.

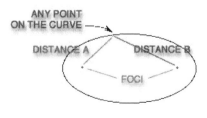

Figure 3.1: Ellipse Foci.

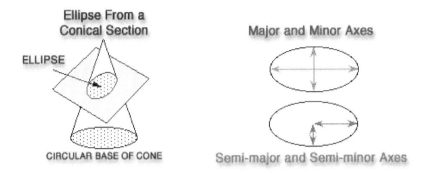

An ellipse results from the intersection of a circular cone and a plane cutting completely though the cone. The maximum diameter is called the major axis. It determines the size of an ellipse. Half the maximum diameter, the distance from the center of the ellipse to one end, is called the semi-major axis.

A more thorough treatment of conic sections, leading to orbital mechanics, is available online.[2]

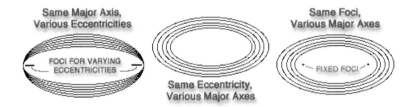

The shape of an ellipse is determined by how close together the foci are in relation to the major axis. Eccentricity is the distance between the foci divided by the major axis. If the foci coincide, the ellipse is a circle. Therefore, a circle is an ellipse with an eccentricity of zero.

Newton's Principles of Mechanics

Sir Isaac Newton realized that the force that makes apples fall to the ground is the same force that makes the planets "fall" around the Sun. Newton had been asked to address the question of why planets move as they do. He established that a force of attraction toward the Sun becomes weaker in proportion to the square of the distance from the Sun.

Newton[3] postulated that the shape of an orbit should be an ellipse. Circular orbits are merely a special case of an ellipse where the foci are coincident. Newton described his work in the Mathematical Principles of Natural Philosophy (often called simply the Principia), which he published in 1685. Newton gave his laws of motion as follows:

1. Every body continues in a state of rest, or of uniform motion in a straight line, unless it is compelled to change that state by forces impressed upon it.

2. The change of motion (linear momentum) is proportional to the force impressed and is made in the direction of the straight line in which that force is impressed.

3. To every action there is always an equal and opposite reaction; or, the mutual actions of two bodies upon each other are always equal, and act in opposite directions.

(Notice that Newton's laws describe the behavior of inertia, they do not explain what the nature of inertia is. This is still a valid question in physics, as is the nature of mass.)

There are three ways to modify the momentum of a body. The mass can be changed, the velocity can be changed (acceleration), or both.

Acceleration

Force (F) equals change in velocity (acceleration, A) times mass (M):

$$F = MA$$

Acceleration may be produced by applying a force to a mass (such as a spacecraft). If applied in the same direction as an object's velocity, the object's velocity increases in relation to an unaccelerated observer. If acceleration is produced by applying a force in the opposite direction from the object's original velocity, it will slow down relative to an unaccelerated observer. If the acceleration is produced by a force at some other angle to the velocity, the object will be deflected. These cases are illustrated below.

The world standard of mass is the kilogram, whose definition is based on the mass of a metal cylinder kept in France. Previously, the standard

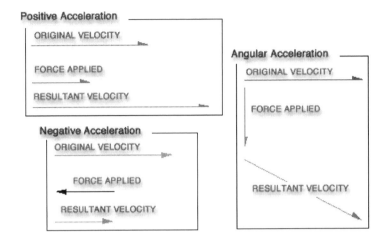

was based upon the mass of one cubic centimeter of water being one gram, which is approximately correct. The standard unit of force is the newton, which is the force required to accelerate a 1-kg mass 1 m/sec^2 (one meter per second per second). A newton is equal to the force from the weight of about 100 g of water in Earth's gravity. That's about half a cup. A dyne is the force required to accelerate a 1 g mass 1 cm/s^2.

The Rocket

A toy water-bottle rocket makes a good example of the basic principle behind how an interplanetary spacecraft, or its launch vehicle, applies Newton's third law. The water-bottle-rocket's energy comes from compressed air, rather than by combustion. The reaction mass is water. When the nozzle is opened, the compressed air accelerates the water out the nozzle, accelerating the rocket forward. Image courtesy Wikimedia Commons.

Figure 3.2: Water Bottle Rocket as an example of the principle of a launch vehicle applying Newton's third law.[4]

A rocket provides the means to accelerate a spacecraft. Like an airplane's jet engine, a rocket creates thrust by expelling mass to take advantage of Sir Isaac Newton's third law (see above). In both systems, combustion increases the temperature of gas in the engine, whereupon it expands and rushes out through a nozzle. The vehicle is accelerated in the opposite direction.

Where a jet engine on an airplane obtains its oxidizer for combustion directly from the atmosphere, a rocket carries its own oxidizer and can operate outside the atmosphere. Otherwise, their basic operating principle is the same.

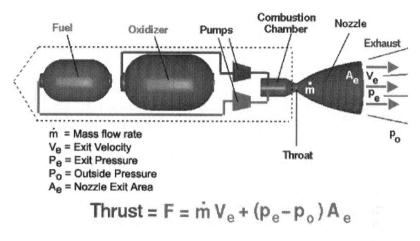

$$\text{Thrust} = F = \dot{m} V_e + (p_e - p_o) A_e$$

Figure 3.3: Schematic of a pumped liquid bipropellant rocket.[5]

The Space Shuttle main engines, and the Vulcain engine in the Ariane 5 launch vehicle, and many others, all use water as the reaction mass, as does a water-bottle rocket. In launch vehicle engines, the water in its gaseous state because it is hot from the combustion process that freshly created it by burning hydrogen with oxygen. This combustion energy (instead of the limited air pressure in a water-bottle rocket) expands and greatly accelerates the reaction mass out the nozzle, propelling the rocket forward.

The reaction mass is not always water. The nature of the exiting mass depends on the chemistry of the propellant(s).[6] Solid-propellant rockets have been used for centuries, operating on the same basic principle, but with different reaction mass and combustion chemistry. Today they are used as strap-on boosters for liquid-fuel launch vehicles, for orbit injection stages, and for many standalone rockets. Modern solid rockets have a core of propellant made of a mixture of fuel and oxidizer in solid form held together by a binder. The solid is molded into a shape having a hollow core that serves as the combustion chamber. Once ignited, a solid rocket cannot be shut off.

The first flight of a liquid-propellant rocket was in 1926 when American professor Robert H. Goddard launched a rocket that used liquid oxygen and gasoline as propellants. It gained 41 feet during a 2.5-second flight. Liquid propellants have the advantage of fairly high density, so the volume and mass of the propellant tanks can be relatively low, resulting in a high mass ratio. Liquid rockets, once started, can be shut off. Some kinds can later be restarted in flight.

An essential part of the modern rocket engine is the convergent-divergent nozzle (CD nozzle) developed by the Swedish inventor Gustaf de Laval in the 19th century who was working to develop a more efficient steam engine. Its operation relies on the different properties of gases flowing at subsonic and supersonic speeds. The speed of a subsonic flow of gas will increase where the conduit carrying it narrows. At subsonic flow the gas is compressible; near the nozzle "throat," where the cross sectional area is a minimum, the gas velocity becomes transonic (Mach number = 1.0), a condition called choked flow. As the nozzle cross-sectional area increases the gas expands, its speed increases to supersonic velocities, and pressure waves cannot propagate backwards. From there, exhaust gas velocity increases at the expense of its pressure and temperature. This nozzle design was key to Goddard, whose engine achieved energy conversion efficiencies of up to 63% — more efficient than any heat-based engine at the time — making space flight a real possibility.

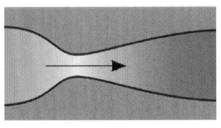

Figure 3.4: DeLaval Nozzle design. Flow velocity increases from left to right.[7]

A thorough discussion of Rocket-Propulsion Basics, including various types of liquid and solid rocket engines, as well as rocket staging, may be found online.[8] The NASA Glenn Research Center offers an interactive nozzle design simulator that allows selection of nozzles for turbine engines or rocket engine, and shows results for various parameter inputs.[9]

A useful parameter that indicates a particular rocket engine's efficiency is its specific impulse, I_{SP}, which is a ratio of the thrust produced to the weight flow of the propellants. It represents the impulse (change in momentum) per unit of propellant. The higher the specific impulse, the less propellant is needed to gain a given amount of momentum. I_{SP} is a useful value to compare engines, much the way "miles per gallon" is used for cars. A propulsion method with a higher specific impulse is more propellant-efficient. I_{SP} is commonly given in seconds (of time) when it is computed using propellant weight (rather than mass). Chapter 11 discusses rocket propulsion systems aboard spacecraft, and Chapter 14 shows a variety of launch vehicles.

Non-Newtonian Physics

We learn from Einstein's special theory of relativity that mass, time, and length are variable, and the speed of light is constant. And from general relativity, we know that gravitation and acceleration are equivalent, that light bends in the presence of mass, and that an accelerating mass radiates gravitational waves at the speed of light.

Figure 3.5: Image of Albert and Elsa Einstein at the Caltech Athenaeum, Pasadena, California, February 1932 courtesy of The Archives, California Institute of Technology. Photo by Anton Schaller.

Spacecraft operate at very high velocities compared to velocities we are familiar with in transportation and ballistics here on our planet. Since spacecraft velocities do not approach a significant fraction of the speed of light, Newtonian physics serves well for operating and navigating throughout the solar system. Nevertheless, accuracies are routinely enhanced by accounting for tiny relativistic effects. Once we begin to travel between the stars, velocities may be large enough fractions of light speed that Einsteinian physics will be indispensable for determining trajectories.

For now, spacecraft do sometimes carry out experiments to test special relativity effects on moving clocks, and experiments to test general relativity effects such as the space-time warp[10] caused by the Sun, frame-dragging,[11] the equivalence of acceleration and gravitation (more precisely the equivalence between inertial mass and gravitational mass)[12] and the search for direct evidence of gravitational waves.[13] As of early 2011 there has been no test by which an observer could tell acceleration from gravitation,[14] nor has gravitational radiation been directly observed. Some of these subjects are explored in Chapter 8.

Acceleration in Orbit

Newton's first law describes how, once in motion, planets remain in motion. What it does not do is explain how the planets are observed to move in nearly circular orbits rather than straight lines. Enter the second law. To move in a curved path, a planet must have an acceleration toward the center of the circle. This is called centripetal acceleration and is supplied by the mutual gravitational attraction between the Sun and the planet.

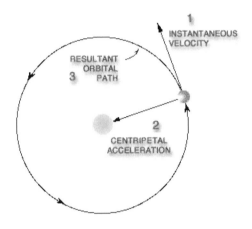

Figure 3.6: Motion and Acceleration in Circular Orbit.

Kepler's Laws

Johannes Kepler's[15] laws of planetary motion are:

1. The orbit of every planet is an ellipse with the Sun at one of the two foci.

2. A line joining a planet and the Sun sweeps out equal areas during equal intervals of time.

3. The square of the orbital period of a planet is directly proportional to the cube of the semi-major axis of its orbit.

The major application of Kepler's first law is to precisely describe the geometric shape of an orbit: an ellipse, unless perturbed by other objects. Kepler's first law also informs us that if a comet, or other object, is observed to have a hyperbolic path, it will visit the Sun only once, unless its encounter with a planet alters its trajectory again.

Kepler's second law addresses the velocity of an object in orbit. Conforming to this law, a comet with a highly elliptical orbit has a velocity at closest approach to the Sun that is many times its velocity when farthest from the Sun. Even so, the area of the orbital plane swept is still constant for any given period of time.

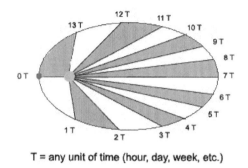

T = any unit of time (hour, day, week, etc.)

Figure 3.7: Illustration of Kepler's Second Law.

Kepler's third law describes the relationship between the masses of two objects mutually revolving around each other and the determination of orbital parameters. Consider a small star in orbit about a more massive one. Both stars actually revolve about a common center of mass, which is called the barycenter. This is true no matter what the size or mass of each of the objects involved. Measuring a star's motion about its barycenter with a massive planet is one method that has been used to discover planetary systems associated with distant stars.

Obviously, these statements apply to a two-dimensional picture of planetary motion, which is all that is needed for describing orbits. A three-dimensional picture of motion would include the path of the Sun through space.

Gravity Gradients & Tidal Forces

Gravity's strength is inversely proportional to the square of the objects' distance from each other. For an object in orbit about a planet, the parts of the object closer to the planet feel a slightly stronger gravitational attraction than do parts on the other side of the object. This is known as gravity gradient. It causes a slight torque to be applied to any orbiting mass which has asymmetric mass distribution (for example, is not spherical), until it assumes a stable attitude with the more massive parts pointing toward the planet. An object whose mass is distributed like a bowling pin would end up in an attitude with its more massive end pointing toward the planet, if all other forces were equal.

Consider the case of a fairly massive body such as our Moon in Earth orbit. The gravity gradient effect has caused the Moon, whose mass is unevenly distributed, to assume a stable rotational rate which keeps one face towards Earth at all times, like the bowling pin described above.

The Moon's gravitation acts upon the Earth's oceans and atmosphere, causing two bulges to form. The bulge on the side of Earth that faces the moon is caused by the proximity of the moon and its relatively stronger gravitational pull on that side. The bulge on the opposite side of Earth results from that side being attracted toward the moon less strongly than is the central part of Earth. Earth's atmosphere and crust are also affected to a smaller degree. Other factors, including Earth's rotation and surface roughness, complicate the tidal effect. On planets or satellites without oceans, the same forces apply, causing slight deformations in the body. This mechanical stress can translate into heat, as in the case of Jupiter's volcanic moon Io.

How Orbits Work

These drawings simplify the physics of orbital mechanics, making it easy to grasp some of the basic concepts. We see Earth with a ridiculously tall mountain rising from it. The mountain, as Isaac Newton first described, has a cannon at its summit.

(1) When the cannon is fired, (Figure 3.8) the cannonball follows its ballistic arc, falling as a result of Earth's gravity, and of course it hits Earth some distance away from the mountain.

(2) If we pack more gunpowder into the cannon, (Figure 3.9) the next time it's fired, the cannonball goes faster and farther away from the mountain, mean-

Figure 3.8: Cannon Atop Impossibly Tall Mountain.

while falling to Earth at the same rate as it did before. The result is that it has gone halfway around the cartoon planet before it hits the ground. (You might enjoy the more elaborate, interactive animation at Space Place.[16])

Figure 3.9: Adding Energy.

In order to make their point these cartoons ignore lots of facts, of course, such as the impossibility of there being such a high mountain on Earth, the drag exerted by the Earth's atmosphere on the cannonball, and the energy a cannon can impart to a projectile... not to mention how hard it would be for climbers to carry everything up such a high mountain! Nevertheless the orbital mechanics they illustrate (in the absence of details like atmosphere) are valid.

(3) Packing still MORE gunpowder into the capable cannon, (Figure 3.10) the cannonball goes much faster, and so much farther that it just never has a chance to touch down. All the while it would be falling to Earth at the same rate as it did in the previous cartoons. This time it falls completely around Earth! We can say it has achieved orbit.

That cannonball would skim past the south pole, and climb right back up to the same altitude from which it was fired, just like the cartoon shows. Its orbit is an ellipse.

This is basically how a spacecraft achieves orbit. It gets an initial boost from a rocket, and then simply falls for the rest of its orbital life. Modern spacecraft are more capable than cannonballs, and they have rocket thrusters

Figure 3.10: Orbital Energy Achieved.

that permit the occasional adjustment in orbit, as described below. Apart from any such rocket engine burns, they're just falling. Launched in 1958 and long silent, the Vanguard-1 Satellite is still falling around Earth in 2011.

In the third cartoon, (Figure 3.10) you'll see that part of the orbit comes closer to Earth's surface that the rest of it does. This is called the periapsis, or periapse, of the orbit. The mountain represents the highest point in the orbit. That's called the apoapsis or apoapse. The altitude affects the time an orbit takes, called the orbit period. The period of the space shuttle's orbit, at say 200 kilometers, is about 90 minutes. Vanguard-1, by the way, has an orbital period of 134.2 minutes, with its periapsis altitude of 654 km, and apoapsis altitude of 3969 km.

The Key to Space Flight

Basically all of space flight involves the following concept, whether orbiting a planet or travelling among the planets while orbiting the Sun.

As you consider the third cartoon, (Figure 3.10) imagine that the cannon has been packed with still more gunpowder, sending the cannonball out a little faster. With this extra energy, the cannonball would miss Earth's surface at periapsis by a greater margin, right?

Right. By applying more energy at apoapsis, you have raised the periapsis altitude.

> **A spacecraft's periapsis altitude can be raised by increasing the spacecraft's energy at apoapsis. This can be accomplished by firing on-board rocket thrusters when at apoapsis.**

And of course, as seen in these cartoons, the opposite is true: if you decrease energy when you're at apoapsis, you'll lower the periapsis altitude. In the cartoons, that's less gunpowder, where the middle graphic (Figure 3.10) shows periapsis low enough to impact the surface. In the next chapter you'll see how this key enables flight from one planet to another.

Now suppose you increase speed when you're at periapsis, by firing an onboard rocket. What would happen to the cannonball in the third cartoon?

Just as you suspect, it will cause the apoapsis altitude to increase. The cannonball would climb to a higher altitude and clear that annoying mountain at apoapsis.

> **A spacecraft's apoapsis altitude can be raised by increasing the spacecraft's energy at periapsis. This can be accomplished by firing on-board rocket thrusters when at periapsis.**

And its opposite is true, too: decreasing energy at periapsis will lower the apoapsis altitude. Imagine the cannonball skimming through the tops of some trees as it flies through periapsis. This slowing effect would rob energy from the cannonball, and it could not continue to climb to quite as high an apoapsis altitude as before.

In practice, you can remove energy from a spacecraft's orbit at periapsis by firing the onboard rocket thrusters there and using up more propellant, or by intentionally and carefully dipping into the planet's atmosphere to use frictional drag. The latter is called aerobraking, a technique used at Venus and at Mars that conserves rocket propellant.

Orbiting a Real Planet

Isaac Newton's cannonball is really a pretty good analogy. It makes it clear that to get a spacecraft into orbit, you need to raise it up and accelerate it until it is going so fast that as it falls, it falls completely around the planet.

In practical terms, you don't generally want to be less than about 150 kilometers above surface of Earth. At that altitude, the atmosphere is so thin that it doesn't present much frictional drag to slow you down. You need your rocket to speed the spacecraft to the neighborhood of 30,000 km/hr (about 19,000 mph). Once you've done that, your spacecraft will continue falling around Earth. No more propulsion is necessary, except for occasional minor adjustments. It can remain in orbit for months or years before the presence of the thin upper atmosphere causes the orbit to degrade. These same mechanical concepts (but different numbers for altitude and speed) apply whether you're talking about orbiting Earth, Venus, Mars, the Moon, the Sun, or anything.

A Periapsis by Any Other Name...

Periapsis and apoapsis, (or periapse and apoapse) are generic terms. The prefixes "peri-" and "ap-" are commonly applied to the Greek or Roman names of the bodies which are being orbited. For example, look for perigee and apogee at Earth, perijove and apojove at Jupiter, periselene and apselene or perilune and apolune in lunar orbit, pericrone and apocrone if you're orbiting Saturn, and perihelion and aphelion if you're orbiting the Sun, and so on.

Freefall

If you ride along with an orbiting spacecraft, you feel as if you are falling, as in fact you are. The condition is properly called freefall. You find yourself falling at the same rate as the spacecraft, which would appear to be floating there (falling) beside you, or around you if you're aboard the International Space Station. You'd just never hit the ground.

Notice that an orbiting spacecraft has not escaped Earth's gravity, which is very much present — it is giving the mass the it needs to stay in orbit. It just happens to be balanced out by the speed that the rocket provided when it placed the spacecraft in orbit. Yes, gravity is a little weaker on orbit, simply because you're farther from Earth's center, but it's mostly there. So terms like "weightless" and "micro gravity" have to be taken with a grain of salt... gravity is still dominant, but some of its familiar effects are not apparent on orbit.

Notes

[1]http://www.astro.ucla.edu/~wright/relatvty.htm
[2]http://www.braeunig.us/space/orbmech.htm
[3]Image of Isaac Newton courtesy newton.org.uk, the Virtual Museum of Sir Isaac Newton.
[4]Image courtesy Wikimedia Commons.
[5]Image courtesy Wikimedia Commons.
[6]http://www.braeunig.us/space/propel.htm
[7]Image courtesy Wikimedia Commons.
[8]http://www.braeunig.us/space/propuls.htm
[9]http://www.grc.nasa.gov/WWW/K-12/airplane/ienzl.html
[10]http://physicsworld.com/cws/article/news/18268
[11]http://einstein.stanford.edu/
[12]http://einstein.stanford.edu/STEP/information/data/general.html
[13]http://www.gothosenterprises.com/gravitational_waves/
[14]http://einstein.stanford.edu/STEP/information/data/gravityhist2.html
[15]http://www-gap.dcs.st-and.ac.uk/~history/Mathematicians/Kepler.html
[16]http://spaceplace.nasa.gov/en/kids/orbits1.shtml

Chapter 4

Interplanetary Trajectories

> **Objectives:** Upon completion of this chapter you will be able to describe the use of Hohmann transfer orbits in general terms and how spacecraft use them for interplanetary travel. You will be able to describe the general concept of exchanging angular momentum between planets and spacecraft to achieve gravity assist trajectories.

When travelling among the planets, it's a good idea to minimize the propellant mass needed by your spacecraft and its launch vehicle. That way, such a flight is possible with current launch capabilities, and costs will not be prohibitive. The amount of propellant needed depends largely on what route you choose. Trajectories that by their nature need a minimum of propellant are therefore of great interest.

Hohmann Transfer Orbits

To launch a spacecraft from Earth to an outer planet such as Mars using the least propellant possible, first consider that the spacecraft is already in solar orbit as it sits on the launch pad. This existing solar orbit must be adjusted to cause it to take the spacecraft to Mars: The desired orbit's perihelion (closest approach to the Sun) will be at the distance of Earth's orbit, and the aphelion (farthest distance from the Sun) will be at the distance of Mars' orbit. This is called a Hohmann Transfer orbit. The portion of the solar orbit that takes the spacecraft from Earth to Mars is called its trajectory.

From the above, we know that the task is to increase the apoapsis (aphelion) of the spacecraft's present solar orbit. Recall from Chapter 3...

> **A spacecraft's apoapsis altitude can be raised by increasing the spacecraft's energy at periapsis.**

Well, the spacecraft is already at periapsis. So the spacecraft lifts off the launch pad, rises above Earth's atmosphere, and uses its rocket to accelerate in the direction of Earth's revolution around the Sun to the extent that the energy added here at periapsis (perihelion) will cause its new orbit to have an aphelion equal to Mars' orbit. The acceleration is tangential to the existing orbit. How much energy to add? See the Rocket & Space Technology website.[1]

After this brief acceleration away from Earth, the spacecraft has achieved its new orbit, and it simply coasts the rest of the way. The launch phase is covered further in Chapter 14.

Earth to Mars via Least Energy Orbit

Getting to the planet Mars, rather than just to its orbit, requires that the spacecraft be inserted into its interplanetary trajectory at the correct time so it will arrive at the Martian orbit when Mars will be there. This task might be compared to throwing a dart at a moving target. You have to lead the aim point by just the right amount to hit the target. The opportunity to launch a spacecraft on a transfer orbit to Mars occurs about every 25 months.

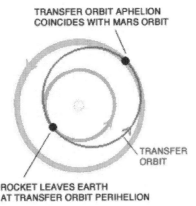

Figure 4.1: Earth to Mars via Least Energy Orbit

To be captured into a Martian orbit, the spacecraft must then decelerate relative to Mars using a retrograde rocket burn or some other means. To land on Mars, the spacecraft must decelerate even further using a retrograde burn to the extent that the lowest point of its Martian orbit will intercept the surface of Mars. Since Mars has an atmosphere, final deceleration may also be performed by aerodynamic braking direct from the interplanetary trajectory, and/or a parachute, and/or further retrograde burns.

Inward Bound

To launch a spacecraft from Earth to an inner planet such as Venus using least propellant, its existing solar orbit (as it sits on the launch pad) must be adjusted so that it will take it to Venus. In other words, the spacecraft's aphelion is already the distance of Earth's orbit, and the perihelion will be on the orbit of Venus.

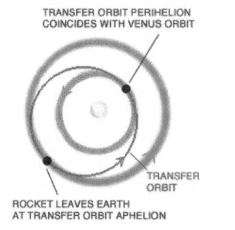

TRANSFER ORBIT PERIHELION
COINCIDES WITH VENUS ORBIT

TRANSFER
ORBIT

ROCKET LEAVES EARTH
AT TRANSFER ORBIT APHELION

Figure 4.2: Earth to Venus via Least Energy Orbit

This time, the task is to decrease the periapsis (perihelion) of the spacecraft's present solar orbit. Recall from Chapter 3...

> **A spacecraft's periapsis altitude can be lowered by decreasing the spacecraft's energy at apoapsis.**

To achieve this, the spacecraft lifts off of the launch pad, rises above Earth's atmosphere, and uses its rocket to accelerate opposite the direction of Earth's revolution around the Sun, thereby decreasing its orbital energy while here at apoapsis (aphelion) to the extent that its new orbit will have a perihelion equal to the distance of Venus's orbit. Once again, the acceleration is tangential to the existing orbit.

Of course the spacecraft will continue going in the same direction as Earth orbits the Sun, but a little slower now. To get to Venus, rather than just to its orbit, again requires that the spacecraft be inserted into its interplanetary trajectory at the correct time so it will arrive at the Venusian orbit when Venus is there. Venus launch opportunities occur about every 19 months.

Type I and II Trajectories

If the interplanetary trajectory carries the spacecraft less than 180 degrees around the Sun, it's called a Type-I Trajectory. If the trajectory carries it 180 degrees or more around the Sun, it's called a Type-II.

Gravity Assist Trajectories

Chapter 1 pointed out that the planets retain most of the solar system's angular momentum. This momentum can be tapped to accelerate spacecraft on so-called "gravity-assist" trajectories. It is commonly stated in the news media that spacecraft such as Voyager, Galileo, and Cassini use a planet's gravity during a flyby to slingshot it farther into space. How does this work? By using gravity to tap into the planet's tremendous angular momentum. In a gravity-assist trajectory, angular momentum is transferred from the orbiting planet to a spacecraft approaching from behind the planet in its progress about the Sun.

> **Note: experimenters and educators may be interested in the Gravity Assist Mechanical Simulator and an illustrated "primer" on gravity assist: `www.jpl.nasa.gov/basics/grav`**

Consider Voyager 2, which toured the Jovian planets. The spacecraft was launched on a Type-II Hohmann transfer orbit to Jupiter. Had Jupiter not been there at the time of the spacecraft's arrival, the spacecraft would have fallen back toward the Sun, and would have remained in elliptical orbit as long as no other forces acted upon it. Perihelion would have been at 1 AU,

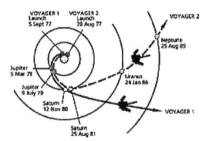

Figure 4.3: Flight Path of Voyager 1 and Voyager 2.

and aphelion at Jupiter's distance of about 5 AU.

However, Voyager's arrival at Jupiter was carefully timed so that it would pass behind Jupiter in its orbit around the Sun. As the spacecraft came into Jupiter's gravitational influence, it fell toward Jupiter, increasing its speed toward maximum at closest approach to Jupiter. Since all masses in the universe attract each other, Jupiter sped up the spacecraft substantially, and the spacecraft tugged on Jupiter, causing the massive planet to actually lose some of its orbital energy.

The spacecraft passed by Jupiter, since Voyager's velocity was greater than Jupiter's escape velocity, and of course it slowed down again relative to Jupiter as it climbed out of the huge gravitational field. The speed component of its Jupiter-relative velocity outbound dropped to the same as that on its inbound leg.

But relative to the Sun, it never slowed all the way to its initial Jupiter approach speed. It left the Jovian environs carrying an increase in angular momentum stolen from Jupiter. Jupiter's gravity served to connect the spacecraft with the planet's ample reserve of angular momentum. This technique was repeated at Saturn and Uranus.

In Figure 4.4, Voyager 2 leaves Earth at about 36 km/s relative to the Sun. Climbing out, it loses much of the initial velocity the launch vehicle provided. Nearing Jupiter, its speed is increased by the planet's gravity, and the spacecraft's velocity exceeds solar system escape velocity. Voyager departs Jupiter with more Sun-relative velocity than it had on arrival. The same is

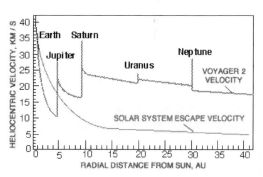

Figure 4.4: Courtesy Steve Matousek, JPL.

seen at Saturn and Uranus. The Neptune flyby design put Voyager close by Neptune's moon Triton rather than attain more speed.

The same can be said of a baseball's acceleration when hit by a bat: angular momentum is transferred from the bat to the slower-moving ball. The bat is slowed down in its "orbit" about the batter, accelerating the ball greatly. The bat connects to the ball not with the force of gravity from behind as was the case with a spacecraft, but with direct mechanical force (electrical force, on the molecular scale, if you prefer) at the front of the bat in its travel about the batter, translating angular momentum from the bat into a high velocity for the ball.

(Of course in the analogy a planet has an attractive force and the bat has a repulsive force, thus Voyager must approach Jupiter from a direc-

tion opposite Jupiter's trajectory and the ball approaches the bat from the
direction of the bats trajectory.)

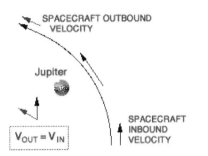

Figure 4.5: Planet-Relative Speeds.

The vector diagram on the left,
Figure 4.5, shows the spacecraft's speed
relative to Jupiter during a gravity-
assist flyby. The spacecraft slows to the
same velocity going away that it had
coming in, relative to Jupiter, although
its direction has changed.   Note also
the temporary increase in speed nearing closest approach.

When the same situation is viewed
as Sun-relative in the diagram below
and to the right, Figure 4.6, we see that Jupiter's Sun-relative orbital velocity is added to the spacecraft's velocity, and the spacecraft does not lose this
component on its way out. Instead, the planet itself loses the energy. The
massive planet's loss is too small to be measured, but the tiny spacecraft's
gain can be very great. Imagine a gnat flying into the path of a speeding
freight train.

Gravity assists can be also used
to decelerate a spacecraft, by flying in
front of a body in its orbit, donating
some of the spacecraft's angular momentum to the body. When the Galileo
spacecraft arrived at Jupiter, passing
close in front of Jupiter's moon Io in
its orbit, Galileo lost energy in relation
to Jupiter, helping it achieve Jupiter
orbit insertion, reducing the propellant
needed for orbit insertion by 90 kg.

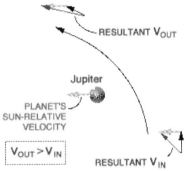

Figure 4.6: Sun-Relative Speeds.

The gravity assist technique was
championed by Michael Minovitch in the early 1960s, while he was a UCLA
graduate student working during the summers at JPL. Prior to the adoption
of the gravity assist technique, it was believed that travel to the outer solar system would only be possible by developing extremely powerful launch
vehicles using nuclear reactors to create tremendous thrust, and basically
flying larger and larger Hohmann transfers.

An interesting fact to consider is that even though a spacecraft may
double its speed as the result of a gravity assist, it feels no acceleration at
all. If you were aboard Voyager 2 when it more than doubled its speed with

gravity assists in the outer solar system, you would feel only a continuous sense of falling. No acceleration. This is due to the virtually equal effect of the planet's gravitation upon all atoms of the passing body.

Enter the Ion Engine

All of the above discussion of interplanetary trajectories is based on the use of today's system of chemical rockets, in which a launch vehicle provides nearly all of the spacecraft's propulsive energy. A few times a year the spacecraft may fire short bursts from its chemical rocket thrusters for small adjustments in trajectory. Otherwise, the spacecraft is in free-fall, coasting all the way

Figure 4.7: Deep Space One.

to its destination. Gravity assists may also provide short periods wherein the spacecraft's trajectory undergoes a change.

But ion electric propulsion, as demonstrated in interplanetary flight by Deep Space 1, works differently. Instead of short bursts of relatively powerful thrust, electric propulsion uses a more gentle thrust continuously over periods of months or even years. It offers a gain in efficiency of an order of magnitude over chemical propulsion for those missions of long enough duration to use the technology. Ion engines are discussed further under Propulsion in Chapter 11.

The Deep Space One Spacecraft[2] used an ion engine. The Japan Aerospace Exploration Agency's asteroid explorer HAYABUSA,[3] which returned samples of dust to Earth from an asteroid, also employs an ion engine.

Even ion-electric propelled spacecraft need to launch using chemical rockets, but because of their efficiency they can be less massive, and require less powerful (and less expensive) launch vehicles. Initially, then, the trajectory of an ion-propelled craft may look like the Hohmann transfer orbit. But over long periods of continuously operating an electric engine, the trajectory will no longer be a purely ballistic arc.

Notes

[1] http://www.braeunig.us/space/orbmech.htm#maneuver
[2] http://nmp.jpl.nasa.gov/ds1
[3] www.jaxa.jp/projects/sat/muses_c/index_e.html

Chapter 5

Planetary Orbits

> **Objectives:** Upon completion of this chapter you will be able to describe in general terms the characteristics of various types of planetary orbits. You will be able to describe the general concepts and advantages of geosynchronous orbits, polar orbits, walking orbits, Sun-synchronous orbits, and some requirements for achieving them.

Orbital Parameters and Elements

The terms orbital period, periapsis, and apoapsis were introduced in Chapter 3. The direction a spacecraft or other body travels in orbit can be direct, or prograde, in which the spacecraft moves in the same direction as the planet rotates, or retrograde, going in a direction opposite the planet's rotation.

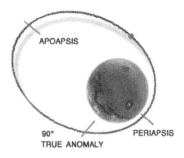

True anomaly is a term used to describe the locations of various points in an orbit. It is the angular distance of a point in an orbit past the point of periapsis, measured in degrees. For example, a spacecraft might cross a planet's equator at 10 true anomaly.

Nodes are points where an orbit crosses a reference plane, such as the ecliptic or the celestial equator. As an orbiting body crosses the reference plane going north, the node is referred to as the ascending node; going south, it is the descending node.

The reference plane to use in defining the ascending node will depend on what you're doing. If you wish to reference your spacecraft's motion to the body it's orbiting, you would use a plane local to that body: orbiting the

Earth you'd use the celestial equator, an extension of the Earth's equatorial plane. Orbiting the Sun you would choose the ecliptic plane, orbiting and studying Venus it would make sense to use an extension of Venus's equatorial plane, etc.

To completely describe an orbit mathematically, six quantities must be calculated. These quantities are called orbital elements, or Keplerian elements, after Johannes Kepler (1571-1630). They are:

1. Semi-major axis and

2. Eccentricity, which together are the basic measurements of the size and shape of the orbit's ellipse (described in Chapter 3. Recall an eccentricity of zero indicates a circular orbit).

3. Inclination is the angular distance of the orbital plane from the plane of the planet's equator (or from the ecliptic plane, if you're talking about heliocentric orbits), stated in degrees. An inclination of 0 degrees means the spacecraft orbits the planet at its equator, and in the same direction as the planet rotates. An inclination of 90 degrees indicates a polar orbit, in which the spacecraft passes over the north and south poles of the planet. An inclination of 180 degrees indicates a retrograde equatorial orbit.

4. Argument of periapsis is the argument (angular distance) of the periapsis from the ascending node.

5. Time of periapsis passage and

6. Celestial longitude of the ascending node are the remaining elements.

The orbital period is of interest to operations, although it is not one of the six Keplerian elements needed to define the orbit.

Generally, three astronomical or radiometric observations of an object in an orbit are enough to pin down all of the above six Keplerian elements. The following table gives a sense of the level of precision an interplanetary mission commonly deals with. These elements are established by analysis of data returned during routine tracking by the Deep Space Network.

Table 5.1: Elements of Magellan's Initial Venus Orbit, 10 August 1990.

1.	Semimajor Axis:	10434.162 km
2.	Eccentricity:	0.2918967
3.	Inclination:	85.69613°
4.	Argument of Periapsis:	170.10651°
5.	Periapsis Passage:	1990 DOY 222 19:54 UTC ERT
6.	Longitude of Ascending Node:	−61.41017°
	(Orbit Period:	3.26375 hr)

Types of Orbits

Geosynchronous Orbits

A geosychronous orbit (GEO) is a prograde, low inclination orbit about Earth having a period of 23 hours 56 minutes 4 seconds. A spacecraft in geosynchronous orbit appears to remain above Earth at a constant longitude, although it may seem to wander north and south. The spacecraft returns to the same point in the sky at the same time each day.

Geostationary Orbits

To achieve a geostationary orbit, a geosychronous orbit is chosen with an eccentricity of zero, and an inclination of either zero, right on the equator, or else low enough that the spacecraft can use propulsive means to constrain the spacecraft's apparent position so it hangs seemingly motionless above a point on Earth. (Any such maneuvering on orbit, or making other adjustments to maintain its orbit, is a process called station keeping.) The orbit can then be called geostationary. This orbit is ideal for certain kinds of communication satellites and meteorological satellites. The idea of a geosynchronous orbit for communications spacecraft was first popularised by science fiction author Sir Arthur C. Clarke in 1945, so it is sometimes called the Clarke orbit.

A Little GTO

To attain geosynchronous (and also geostationary) Earth orbits, a spacecraft is first launched into an elliptical orbit with an apoapsis altitude in the neighborhood of 37,000 km. This is called a Geosynchronous Transfer Orbit (GTO). The spacecraft, or a separable upper stage, then circularizes the orbit by turning parallel to the equator at apoapsis and firing its rocket engine. That engine is usually called an apogee motor. It is common to compare various launch vehicles' capabilities (see Chapter 14) according to the amount of mass they can lift to GTO.

Polar Orbits

Polar orbits are 90-degree inclination orbits, useful for spacecraft that carry out mapping or surveillance operations. Since the orbital plane is nominally fixed in inertial space, the planet rotates below a polar orbit, allowing the spacecraft low-altitude access to virtually every point on the surface. The Magellan spacecraft used a nearly-polar orbit at Venus. Each periapsis pass, a swath of mapping data was taken, and the planet rotated so that swaths from consecutive orbits were adjacent to each other. When the planet rotated once, all 360 degrees longitude had been exposed to Magellan's surveillance.

To achieve a polar orbit at Earth requires more energy, thus more propellant, than does a direct (prograde) orbit of low inclination. To achieve the latter, launch is normally accomplished near the equator, where the rotational speed of the surface contributes a significant part of the final speed required for orbit. A polar orbit will not be able to take advantage of the "free ride" provided by Earth's rotation, and thus the launch vehicle must provide all of the energy for attaining orbital speed.

Walking Orbits

Planets are not perfectly spherical, and they do not have evenly distributed surface mass. Also, they do not exist in a gravity "vacuum." Other bodies such as the Sun, or natural satellites, contribute their gravitational influences to a spacecraft in orbit about a planet. It is possible to choose the parameters of a spacecraft's orbit to take advantage of some or all of these gravitational influences to induce precession, which causes a useful motion of the orbital plane. The result is called a walking orbit or a precessing orbit, since the orbital plane moves slowly with respect to fixed inertial space.

Sun Synchronous Orbits

A walking orbit whose parameters are chosen such that the orbital plane precesses with nearly the same period as the planet's solar orbit period is called a Sun synchronous orbit. In such an orbit, the spacecraft crosses periapsis at about the same local time every orbit. This can be useful if instruments on board depend on a certain angle of solar illumination on the surface. Mars Global Surveyor's orbit was a 2-pm Mars Local Time Sun-synchronous orbit, chosen to permit well-placed shadows for best viewing.

It may not be possible to rely on use of the gravity field alone to exactly maintain a desired synchronous timing, and occasional propulsive maneuvers may be necessary to adjust the orbit.

Lagrange points and "Halo" Orbits

Joseph Louis Lagrange (1736-1813) showed how a body of negligible mass (commonly called a "particle," but this can include a spacecraft) could orbit along with a larger body that is already in a near-circular orbit. In a frame of reference that rotates with the larger bodies, he found five points at which the combined gravitational forces of the two larger bodies can maintain a particle rotating in a constant relative position as they orbit.

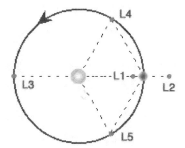

Figure 5.1: Lagrange Points.

Consider a system with the two large bodies being the Earth orbiting the Sun. The third body, such as a spacecraft, might occupy any of five Lagrangian points:

In line with the two large bodies are the L1, L2 and L3 points, (see Figure 5.1) which are unstable: a spacecraft located there must use its thrusters occasionally to remain near the point.

The leading apex of the triangle is L4; the trailing apex is L5. These last two are also called Trojan points. Natural bodies such as asteroids can often be found at a planet's L4 and/or L5. Trojan moons may also be found orbiting a planet at the planet-moon L4 and L5 points.

With a minimum use of thrusters for stationkeeping, a spacecraft can "orbit" about an unstable Lagrange point. Such an orbit is called a halo orbit because it appears as an ellipse floating over the planet. It is not an orbit in the classical sense, though, since the unstable L point does not exert any attractive force on its own. In the Sun-Earth case for example, the spacecraft's true orbit is around the Sun, with a period equal to Earth's (the year). Picture a halo orbit as a controlled drift back and forth in the vicinity of the L point while orbiting the Sun. Most of the propulsive stationkeeping maneuvers are executed out near the extremities, or ansas, of the halo, reversing the direction of drift each time by gentle force.

The Solar and Heliospheric Observatory spacecraft (SOHO)[1] follows a "halo" orbit around Earth's L1, which is 1.53×10^6 km from Earth. From there it has an uninterrupted view of its target the Sun, and its orbital excursions ensure that ground stations are not always pointing right at the noisy Sun for communications. The Wilkinson Microwave Anisotropy Probe (WMAP)[2] spacecraft resides in a halo orbit near Earth's L2 (about the same distance from Earth as L1), where it enjoys an uninterrupted view into deep space. Its 6-month orbit about L2 prevents Earth's shadow from ever blocking the craft's solar arrays. An online movie[3] of WMAP's flight illustrates the halo orbit.

Notes

[1]http://sohowww.nascom.nasa.gov
[2]http://map.gsfc.nasa.gov
[3]http://map.gsfc.nasa.gov/m_ig/anim009/anim009/L2.avi

Chapter 6

Electromagnetic Phenomena

> **Objectives:** Upon completion of this chapter you will be able to describe in
> general terms characteristics of natural and artificial emitters of radiation.
> You will be able to describe bands of the spectrum from RF to gamma rays,
> and the particular usefulness radio frequencies have for deep-space commu-
> nication. You will be able to describe the basic principles of spectroscopy,
> Doppler effect, reflection and refraction.

Electromagnetic Radiation

Electromagnetic radiation (radio waves, light, etc.)
consists of interacting, self-sustaining electric and
magnetic fields that propagate through empty space
at 299,792 km per second (the speed of light, c),[2] and
slightly slower through air and other media. Thermonuclear reactions in the cores of stars (including
the Sun) provide the energy that eventually leaves
stars, primarily in the form of electromagnetic radi-

Figure 6.1: M45.[1]

ation. These waves cover a wide spectrum of frequencies. Sunlight is a
familiar example of electromagnetic radiation that is naturally emitted by
the Sun. Starlight is the same thing from "suns" much farther away.

When a direct current (DC) of electricity, for example from a flashlight
battery, is applied to a wire or other conductor, the current flow builds an
electromagnetic field around the wire, propagating a wave outward. When
the current is removed the field collapses, again propagating a wave. If
the current is applied and removed repeatedly over a period of time, or
if the electrical current is made to alternate its polarity with a uniform
period of time, a series of waves is propagated at a discrete frequency. This

phenomenon is the basis of electromagnetic radiation.

Electromagnetic radiation normally propagates in straight lines at the speed of light and does not require a medium for transmission. It slows as it passes through a medium such as air, water, glass, etc.

The Inverse Square Law

Electromagnetic energy decreases as if it were dispersed over the area on an expanding sphere, expressed as $4\pi R^2$ where radius R is the distance the energy has travelled. The amount of energy received at a point on that sphere diminishes as $1/R^2$. This relationship is known as the inverse-square law of (electromagnetic) propagation. It accounts for loss of signal strength over space, called space loss.

The inverse-square law is signif-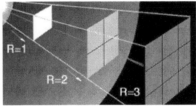
icant to the exploration of the uni-
verse, because it means that the con-
centration of electromagnetic radiation
decreases very rapidly with increasing
distance from the emitter. Whether
the emitter is a distant spacecraft with a low-power transmitter or an extremely powerful star, it will deliver only a small amount of electromagnetic energy to a detector on Earth because of the very great distances and the small area that Earth subtends on the huge imaginary sphere.

Electromagnetic Spectrum

Light is electromagnetic radiation (or electromagnetic force) at frequencies that can be sensed by the human eye. The whole electromagnetic spectrum, though, has a much broader range of frequencies than the human eye can detect, including, in order of increasing frequency: audio frequency (AF), radio frequency (RF), infrared (meaning "below red," IR), visible light, ultraviolet (meaning "above violet," UV), X-rays, and finally gamma rays. These designations describe only different frequencies of the same phenomenon: electromagnetic radiation.

All electromagnetic waves propagate at the speed of light. The wavelength of a single oscillation of electromagnetic radiation means the distance the wave will propagate *in vacuo* during the time required for one oscillation. Frequency is expressed in hertz (Hz), which represents cycles per second.

The strength, or intensity of the wave (analogous to a sound wave's loudness) is known as its amplitude.

There is a simple relationship between the frequency of oscillation and wavelength of electromagnetic energy. Wavelength, represented by the Greek lower case lambda (λ), is equal to the speed of light (c) divided by frequency (f).

$$\lambda = c/f$$

and

$$f = c/\lambda$$

dB - Power Ratios in Decibels

For wavelengths up through radio frequencies, power or intensity is commonly expressed as a power ratio in decibels, dB. This is a means of convenient notation (as is also the system of scientific notation) to describe power ratios varying over many orders of magnitude. The decibel takes the form of a base-10 logarithm. A reference may be specified, for example, dBm is referenced to milliwatts, dBw is referenced to watts, etc. Example:

$$20 \ dBm = 10^{20/10} = 10^2 = 100 \ milliwatts$$

An increase of 3 dB corresponds to an approximate doubling of power. The factor is specifically $10^{3/10} = 1.9953$.

Waves or Particles?

Electromagnetic radiation of all frequencies or energies can be viewed in physics as if it were waves, as described above, and also as particles, known as photons. It is generally common to speak of waves when talking about lower frequencies and longer wavelengths, such as radio waves. Reference to photons is common for physicists talking about light and electromagnetic force of higher frequencies (or energies). Waves are described in terms of frequency, wavelength, and amplitude. Photons, seen as particle carriers of the electromagnetic force, are described in terms of energy level using the electron volt (eV). Throughout this document the preferred treatment will be waves, which is arguably a more informative approach.

Natural and Artificial Emitters

Deep space communication antennas and receivers are capable of detecting many different kinds of natural emitters of electromagnetic radiation, including the stars, the Sun, molecular clouds, and gas giant planets such as Jupiter. These sources do not emit at truly random frequencies, but without sophisticated scientific investigation and research, their signals appear as noise – that is, signals of pseudo-random frequencies and amplitudes. Radio Astronomy[3] is

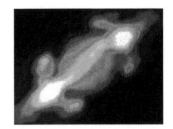

Figure 6.2: Radio Image of Jupiter, courtesy National Radio Astronomical Observatory.

the scientific discipline which investigates natural emitters by acquiring and studying their electromagnetic radiation. The Deep Space Network participates in radio astronomy experiments.

Deep space vehicles are equipped with radio transmitters ("artificial emitters") and receivers for sending and receiving signals (electromagnetic radiation) to and from Earth-based tracking stations. These signals utilize pre-established discrete frequencies. On the other hand, various natural and human-made emitters combine to create a background of electromagnetic noise from which the spacecraft signals must be detected. The ratio of the signal level to the noise level is known as the signal-to-noise ratio (SNR). SNR is commonly expressed in decibels.

Radio Frequencies

Abbreviations such as kHz and GHz are all listed in the Glossary and are also treated under Units of Measure.

Electromagnetic radiation with frequencies between about 10 kHz and 100 GHz are referred to as radio frequencies (RF). Radio frequencies are divided into groups that have similar characteristics, called "bands," such as "S-band," "X-band," etc. The bands are further divided into small ranges of frequencies called "channels," some of which are allocated for the use of deep space telecommunications.

Many deep-space vehicles use channels in the S-band and X-band range which are in the neighborhood of 2 to 10 GHz. These frequencies are among those referred to as microwaves, because their wavelength is short, on the order of centimeters. The microwave oven takes its name from this range of radio frequencies. Deep space telecommunications systems are now being

Table 6.1: Radio Frequency Bands.

Band	Approx. Range of Wavelengths (cm)	Approximate Frequencies
UHF	100 – 10	300 – 3000 MHz
L	30 – 15	1 – 2 GHz
S	15 – 7.5	2 – 4 GHz
C	7.5 – 3.75	4 – 8 GHz
X	3.75 – 2.4	8 – 12 GHz
K*	2.4 – 0.75	12 – 40 GHz
Q	0.75 – 0.6	40 – 50 GHz
V	0.6 – 0.4	50 – 80 GHz
W	0.4 – 0.3	80 – 90 GHz

*Within K-band, spacecraft may operate communications, radio science, or radar equipment at Ku-band in the neighborhood of 15 to 17 GHz and Ka-band around 20 to 30 GHz.

operated on the even higher frequency K-band.

This table lists some RF band definitions for illustration. Band definitions may vary among different sources and according to various users. These are "ballpark" values, intended to offer perspective, since band definitions have not evolved to follow any simple alphabetical sequence. For example, notice that while "L-Band" represents lower frequencies than "S-Band," "Q-Band" represents higher frequencies than "S-Band."

The Whole Spectrum

Table 6.2 (page 95) illustrates the entire electromagnetic spectrum in terms of frequency and wavelength (though it is more conventional to speak of ultraviolet light, X-rays, and gamma rays in units of photon energy using the electron volt, eV). It shows common names such as "visible light" and "gamma rays," size examples, and any attenuation effects in Earth's environment as discussed below. The following abbreviations appear in the table: for frequency, THz = Terahertz, PHz = Petahertz, EHz = Exahertz, ZHz = Zettahertz, YHz = Yottahertz; for wavelength, μ = micrometer or micron, nm = nanometer, pm = picometer, fm = femtometer, and am = attometer. The Angstrom, Å, is 0.1 nm. Angsrtoms have traditionally been used to describe wavelengths of light; the nanometer is generally preferred today. (Note: infrared extends all the way to red light, and ultraviolet extends from blue, despite the coarser granularity used for convenience.)

Atmospheric Transparency

Because of the absorption phenomenon, observations are impossible at certain wavelengths from the surface of Earth, since they are absorbed by the Earth's atmosphere. There are a few "windows" in its absorption characteristics that make it possible to see visible light and receive many radio frequencies, for example, but the atmosphere presents an opaque barrier to much of the electromagnetic spectrum.

Even though the atmosphere is transparent at X-band frequencies, there is a problem when liquid or solid water is present. Water exhibits noise at X-band frequencies, so precipitation at a receiving site increases the system noise temperature, and this can drive the SNR too low to permit communications reception.

Radio Frequency Interference

In addition to the natural interference that comes from water at X-band, there may be other sources of noise, such as human-made radio interference. Welding operations on an antenna produce a wide spectrum of radio noise at close proximity to the receiver. Many Earth-orbiting spacecraft have strong downlinks near the frequency of signals from deep space. Goldstone Solar System Radar (described further in this chapter) uses a very powerful transmitter, which can interfere with reception at a nearby station. Whatever the source of radio frequency interference (RFI), its effect is to increase the noise, thereby decreasing the SNR and making it more difficult, or impossible, to receive valid data from a deep-space craft.

Spectroscopy

The study of the production, measurement, and interpretation of electromagnetic spectra is known as spectroscopy. This branch of science pertains to space exploration in many different ways. It can provide such diverse information as the chemical composition of an object, the speed of an object's travel, its temperature, and more – information that cannot be gleaned from photographs or other means.

Table 6.2: Characteristics of Electromagnetic Energy

Band	Frequency	Wavelength	Size Example	Attenuation
Long-wave radio	3 kHz	100 km		Ionosphere opaque
	3×10^3 Hz	100×10^3 m	Los Angeles	↓
	30 kHz	10 km		
	30×10^3 Hz	10×10^3 m	Pasadena	↓
AM radio	300 kHz	1 km		
	300×10^3 Hz	1×10^3 m	JPL	↓
Short-wave radio	3 MHz	100 m		
	3×10^6 Hz		Football field	↓
VHF radio, FM, TV	30 MHz	10 m		
	30×10^6 Hz		Conference room	↓
UHF radio, TV	300 MHz	1 m		
	300×10^6 Hz		Human child	↓
Microwave radio	3 GHz	100 mm		Ionosphere opaque
	3×10^9 Hz	100×10^{-3} m	Cellphone	
	30 GHz	10 mm		Atmosphere opaque
	30×10^9 Hz	10×10^{-3} m	Pebble	at most wavelengths
	300 GHz	1 mm		↓
	300×10^9 Hz	1×10^{-3} m	Grain of sand	
Infrared light	3 THz	100 μ		↓
	3×10^{12} Hz	100×10^{-6} m	Dust mite	
	30 THz	10 μ		↓
	30×10^{12} Hz	10×10^{-6} m	Pollen grain	
	300 THz	1 μ		Atmosphere opaque
	300×10^{12} Hz	1×10^{-6} m	Bacterium	at most wavelengths
Visible light	▬▬▬	700 – 400 nm	Virus	Atmosphere transparant
		7,000 – 4,000 Å		
Ultraviolet light	3 PHz	100 nm		Atmosphere opaque
	3×10^{15} Hz	100×10^{-9} m	Virus	
	30 PHz	10 nm		
	30×10^{15} Hz	10×10^{-9} m		↓
	300 PHz	1 nm		
	300×10^{15} Hz	1×10^{-9} m		↓
X-rays	3 EHz	100 pm		
	3×10^{18} Hz	100×10^{-12} m	Atom	↓
	30 EHz	10 pm		
	30×10^{18} Hz	10×10^{-12} m		↓
	300 EHz	1 pm		
	300×10^{18} Hz	1×10^{-12} m		↓
Gamma rays	3 ZHz	100 fm		
	3×10^{21} Hz	100×10^{-15} m		↓
	30 ZHz	10 fm		
	30×10^{21} Hz	10×10^{-15} m		↓
	300 ZHz	1 fm		
	300×10^{21} Hz	1×10^{-15} m	Atomic nucleus	↓
	300 YHz	1 am		
	300×10^{24} Hz	1×10^{-18} m		Atmosphere opaque

For purposes of introduction, imagine sunlight passing through a glass prism, creating a rainbow, called the spectrum. Each band of color visible in this spectrum is actually composed of a very large number of individual wavelengths of light which cannot be individually discerned by the human eye, but which are detectable by sensitive instruments such as spectrometers and spectrographs.

Suppose instead of green all you find is a dark "line" where green should be. You might assume something had absorbed all the "green" wavelengths out of the incoming light. This can happen. By studying the brightness of individual wavelengths from a natural source, and comparing them to the results of laboratory experiments, many substances can be identified that lie in the path from the light source to the observer, each absorbing particular wavelengths, in a characteristic manner.

Figure 6.3: Dark Absorption Lines in the Sun's Spectrum

Dark absorption lines in the Sun's spectrum and that of other stars are called Fraunhofer lines after Joseph von Fraunhofer (1787-1826) who observed them in 1817. Figure 6.3 shows a segment of the solar spectrum, in which many such lines can be seen. The prominent dark line above the arrow results from hydrogen in the Sun's atmosphere absorbing energy at a wavelength of 6563 Angstroms. This is called the hydrogen alpha line.

On the other hand, bright lines in a spectrum (not illustrated here) represent a particularly strong emission of radiation produced by the source at a particular wavelength.

Spectroscopy is not limited to the band of visible light, but is commonly applied to infrared, ultraviolet, and many other parts of the whole spectrum of electromagnetic energy.

In 1859, Gustav Kirchhoff (1824-1887) described three laws of spectral analysis:

1. A luminous (glowing) solid or liquid emits light of all wavelengths (white light), thus producing a continuous spectrum.

2. A rarefied luminous gas emits light whose spectrum shows bright lines (indicating light at specific wavelengths), and sometimes a faint superimposed continuous spectrum.

3. If the white light from a luminous source is passed through a gas, the gas may absorb certain wavelengths from the continuous spectrum so that those wavelengths will be missing or diminished in its spectrum, thus producing dark lines.

By studying emission and absorption features in the spectra of stars, in the spectra of sunlight reflected off the surfaces of planets, rings, and satellites, and in the spectra of starlight passing through planetary atmospheres, much can be learned about these bodies. This is why spectral instruments are flown on spacecraft.

Historically, spectral observations have taken the form of photographic prints showing spectral bands with light and dark lines. Modern instruments (discussed again under Chapter 12) produce their high-resolution results in the form of X-Y graphic plots, whose peaks and valleys reveal intensity (brightness) on the vertical axis versus wavelength along the horizontal. Peaks of high intensity on such a plot represent bright spectral lines (not seen in Figure 6.3), and troughs of low intensity represent the dark lines.

The plot in Figure 6.4, reproduced courtesy of the *Institut National des Sciences de l'Univers / Observatoire de Paris,* shows details surrounding the dip in solar brightness centered at the hydrogen-alpha line of 6563 Å (angstroms) which is indicated by the dark line above the red arrow in the spectral image in Figure 6.3. The whole plot spans 25 Å of wavelength horizontally.

Spectral observations of distant supernovae (exploding stars) provide data for astrophysicists to understand the supernova process, and to categorize the various supernova types. Supernovae can be occasionally found in extremely distant galaxies. Recognizing their spectral signature is an important step in measuring the size of the universe, based on knowing the original brightness of a supernova and comparing that with the observed brightness across the distance.

The Doppler Effect

Regardless of the frequency of a source of electromagnetic waves, they are subject to the Doppler effect. The effect was discovered by the Austrian mathematician and physicist Christian Doppler (1803-1853). It causes the observed frequency of any source (sound, radio, light, etc.) to differ

Christian Doppler

from the radiated frequency of the source if there is motion that is increasing or decreasing the distance between the source and the observer. The effect

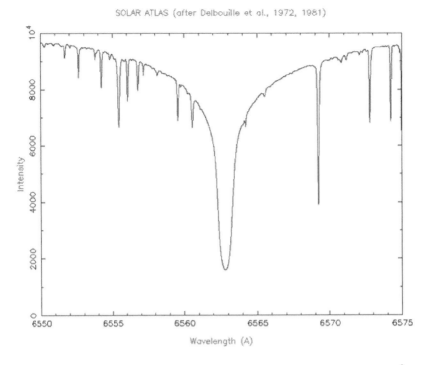

Figure 6.4: Plot of Hydrogen-alpha Solar Absorption around 6563 Å.

is readily observable as variation in the pitch of sound[4] between a moving source and a stationary observer, or vice-versa.

Consider the following:

1. When the distance between the source and receiver of electromagnetic waves remains constant, the frequency of the source and received wave forms is the same.

2. When the distance between the source and receiver of electromagnetic waves is increasing, the frequency of the received wave forms appears to be lower than the actual frequency of the source wave form. Each time the source has completed a wave, it has also moved farther away from the receiver, so the waves arrive less frequently.

3. When the distance is decreasing, the frequency of the received wave form will be higher than the source wave form. Since the source is getting closer, the waves arrive more frequently.

The Doppler effect is routinely measured in the frequency of the signals received by ground receiving stations when tracking spacecraft. The increasing or decreasing distances between the spacecraft and the ground station may be caused by a combination of the spacecraft's trajectory, its orbit around a planet, Earth's revolution about the Sun, and Earth's daily rotation on its axis. A spacecraft approaching Earth will add a positive frequency bias to the received signal. However, if it flies by Earth, the received Doppler bias will become zero as it passes Earth, and then become negative as the spacecraft moves away from Earth.

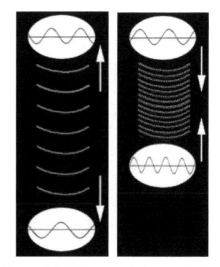

Figure 6.5: Relative radial motion induces Doppler shift.

A spacecraft's revolutions around another planet such as Mars adds alternating positive and negative frequency biases to the received signal, as the spacecraft first moves toward and then away from Earth.

Differenced Doppler

The Earth's daily rotation adds a positive frequency bias to the received signal as the spacecraft rises in the east at a particular tracking station, and it adds a negative frequency bias to the received signal as the spacecraft sets in the west.

The Earth's revolution about the Sun adds a positive frequency bias to the received signal during that portion of the year when the Earth is moving toward the spacecraft, and it adds a negative frequency bias during the part of the year when the Earth is moving away.

Figure 6.6: Differenced Doppler.

If two widely-separated tracking stations on Earth observe a single spacecraft in orbit about another planet, they will each have a slightly different view of the moving spacecraft, and there will be a slight difference in the amount of Doppler shift observed by each station. For example, if one station has a view exactly edge-on to the spacecraft's orbital plane, the other station would have a view slightly to one side of that plane. Information can be extracted from the differencing of the two received signals.

Data obtained from two stations in this way can be combined and interpreted to fully describe the spacecraft's arc through space in three dimensions, rather than just providing a single toward or away component. This data type, differenced Doppler, is a useful form of navigation data that can yield a very high degree of spatial resolution. It is further discussed in Chapter 13, Spacecraft Navigation.

Reflection

Electromagnetic radiation travels through empty space in a straight line except when it is bent slightly by the gravitational field of a large mass in accordance with general relativity.

Radio Frequency (RF) waves can be reflected by certain substances, in much the same way that light is reflected by a mirror. As with light on a mirror, the angle at which RF is reflected from a smooth metal surface, for example, will equal the angle at which it approached the surface.

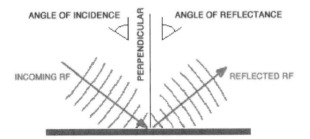

Figure 6.7: The reflectance angle of RF waves equals their incidence angle.

The principle of RF reflectionis used in designing antennas to focus incoming microwave radio energy from a large area down into a narrow beam, collecting and concentrating it into a receiver. If a reflector is shaped like a paraboloid, RF electromagnetic waves approaching on-axis (and only those) will reflect and focus at the feed horn.

This arrangement, called prime focus, makes use of a large reflector to receive very weak RF signals. It is also used in optical telescopes. But prime focus arrangements for large radio antennas place heavy prime-focus equipment far from the main reflector, so the supporting structure tends to sag under its own weight, affecting the system's ability to focus. It also exposes large structures to the wind.

A solution is the Cassegrain focus arrangement. Cassegrain antennas add a secondary reflecting surface, called a subreflector, to "fold" the RF back to a focus near the primary reflector. All the DSN's antennas are of this design because it accommodates large apertures and is structurally strong, allowing heavy equipment to be located nearer the structure's center of gravity. More information on the design of DSN stations appears in Chapter 18. Many optical telescopes both large and small also use Cassegrain or similar systems.

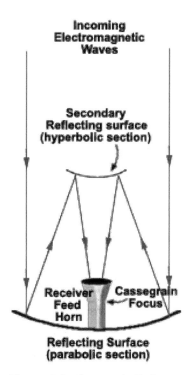

Figure 6.8: Cassegrain Reflector.

Planetary Radar

The reflective properties of RF electromagnetic waves have also been used to investigate the planets and their satellites using a technique called planetary radar astronomy. With this technique, electromagnetic waves are radiated from high-power transmitters in the DSN antennas and they reflect off the surface of the planet or satellite, to be received at one or more Earth stations. Using very sophisticated signal processing techniques, the receiving stations dissect and analyze the signal in terms of time, amplitude, phase, and frequency.

JPL's application of this radar technique, called Goldstone Solar System Radar (GSSR), has been used to develop images of the surface features of Venus, which is eternally covered with clouds, of Mercury, difficult to see visually in the glare of the Sun, and of satellites of the Jovian planets, including Saturn's large moon Titan, whose surface is obscured from view by a thick hazy atmosphere.

The Arecebo Radio Telescope offers good examples of RF reflecting schemes. This 305-meter aperture facility serves for both radio astronomy and radar astronomy investigations.

Reflection of X-rays

At much shorter wavelength and higher energy than RF and light, X-rays do not reflect in the same manner at the large incidence angles employed in optical and
radio telescopes. Instead, they are mostly absorbed. To bring X-rays to a focus, one has to use a different approach from Cassegrain or other typical reflecting systems.

Much like skipping a stone on the water by throwing it at a low angle to the surface (experiments reveal that 20° to the water is best!), X-rays may be deflected by mirrors arranged at low incidence angles to the incoming energy. Mirrors in X-ray and gamma-ray telescopes are arrangements of flat surfaces or concentric tubes configured so that the incoming X-rays strike at glancing (grazing) incidence and deflect toward a focal point in one or more stages.

The Chandra X-ray Observatory, one of NASA's four "Great Observatories," has been operating in Earth orbit since August 1999. Chandra uses two sets of 4 nested grazing-incidence mirrors to bring X-ray photons to focus onto detector instruments as illustrated in Figure 6.9. The Chandra website[5] offers several animations that illustrate the concept.

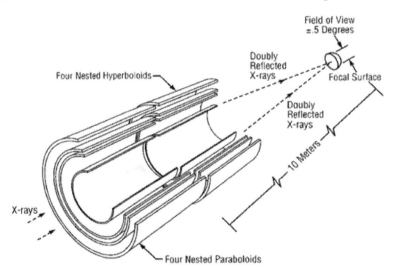

Figure 6.9: Nested glancing-incidence mirrors of the Chandra X-ray Observatory. Image courtesy NASA/CXC/SAO.

Refraction

Refraction is the deflection or bending of electromagnetic waves when they pass from one kind of transparent medium into another. The index of refraction of a material is the ratio of the speed of light in a vacuum to the speed of light in the material. Electromagnetic waves passing from one medium into another of a differing index of refraction will be bent in their direction of travel. In 1621, Dutch physicist Willebrord Snell (1591-1626), determined the angular relationships of light passing from one transparent medium to another.[6]

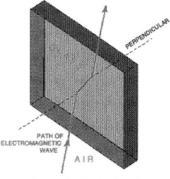

Air and glass have different indices of refraction. Therefore, the path of electromagnetic waves moving from air to glass at an angle will be bent toward the perpendicular as they travel into the glass. Likewise, the path will be bent to the same extent away from the perpendicular when they exit the other side of glass.

Refraction is responsible for many useful devices which bend light in carefully determined ways, from eyeglasses to refracting telescope lenses.[7]

Figure 6.10: Refraction

Refraction can cause illusions. The pencil below left appears to be discontinuous at the boundary of air and water. Likewise, spacecraft may appear to be in different locations in the sky than they really are.

Electromagnetic waves entering Earth's atmosphere from space are bent by refraction. Atmospheric refraction is greatest for signals near the horizon where they come in at the lowest angle. The apparent altitude of the signal source can be on the order of half a degree higher than its true height (see Figure 6.11).

As Earth rotates and the object gains altitude, the refraction effect reduces, becoming zero at the zenith (directly overhead). Refraction's effect on the Sun adds about 5 minutes of time to the daylight at equatorial latitudes, since it appears higher in the sky than it actually is.

If the signal from a spacecraft goes through the atmosphere of another planet, the signals leaving the spacecraft will be bent by the atmosphere of that planet. This bending will cause the apparent occultation, that is, going behind the planet, to occur later than otherwise expected, and to exit from occultation prior to when otherwise expected. Ground processing of the received signals reveals the extent of atmospheric bending, and also of absorption at specific frequencies and other modifications. These provide a basis for inferring the composition and structure of a planet's atmosphere.

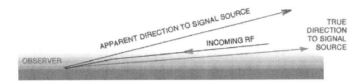

Figure 6.11: Refraction in Earth's Atmosphere (angles exaggerated for clarity).

Phase

As applied to waves of electromagnetic radiation, phase is the relative measure of the alignment between two waveforms of similar frequency. They are said to be in phase if the peaks and troughs of the two waves match up with each other in time. They are said to be out of phase to the extent that they do not match up. Phase is expressed in degrees from 0 to 360. See Figure 6.12.

Figure 6.12: Waves in phase, 180° out of phase, and 40° out of phase.

The phase difference between two waves, or phase change in a single wave, can apply in different areas of interplanetary space flight. Phase shifts in a spacecraft's telecommunications radio signal can be employed to carry information. Observations of effects from the Sun (or other body) upon the phase (and other characteristics) of a radio signal can provide information about that body to radio scientists. Interactions between radio waves, or light waves, given their phase relationships, can also be useful in many ways.

Wave Interactions

Waves can interact with one another in various ways. For example, two incoming radio waves can augment one another, if they are in phase. If they are out of phase they can cancel each other out. Such interactions, and many other kinds, are involved variously in the business of interplanetary space flight: from scientific instruments aboard spacecraft that employ interferometers, [8] to tracking antennas arrayed (employing interferometry) to increase the power of a spacecraft's radio signal being received, or ground-based telescopes that use interferometry to achieve enormous gains in resolution.

No further discussion of the physics of wave interaction is included here, but the subject can readily be researched via resources on the internet.

Notes

[1]Image of M45, The Pleiades star cluster, by kind permission of Graham Pattison who acquired the image.

[2]http://www.speed-light.info

[3]http://www.jpl.nasa.gov/radioastronomy

[4]http://paws.kettering.edu/~drussell/Demos/doppler/carhorn.wav

[5]http://chandra.harvard.edu/resources/animations/mirror.html

[6]http://interactagram.com/physics/optics/refraction

[7]http://www.meade.com/support/telewrk.html

[8]http://planetquest.jpl.nasa.gov/Keck

Part II

FLIGHT PROJECTS

Chapter 7

Mission Inception Overview

> **Objectives:** Upon completion of this chapter you will be able to describe activities typical of the following mission phases: conceptual effort, preliminary analysis, definition, design, and development. You will be conversant with typical design considerations included in mission inception.

In this discussion, we will consider science projects suitable for sponsorship by the U.S. National Aeronautics and Space Administration (NASA). While a large percentage of projects at JPL enjoy sponsorship by NASA, many JPL projects do have different sponsors. This discussion considers a hypothetical example, and it is effective in offering a valid basis for comparison to real-world projects. In reality though, there may be many deviations from this nominal process.

There is no single avenue by which a mission must be initiated. An original concept for a mission to obtain scientific data may come from members of the science community who are interested in particular aspects of certain solar system bodies, or it may come from an individual or group, such as a navigation team, who know of a unique opportunity approaching from an astronomical viewpoint. As a project matures, the effort typically goes through these different phases:

1. Pre-Phase A, Conceptual Study

2. Phase A, Preliminary Analysis

3. Phase B, Definition

4. Phase C/D, Design and Development

5. Phase E, Operations Phase

Formal reviews are typically used as control gates at critical points in the full system life cycle to determine whether the system development process should continue from one phase to the next, or what modifications may be required.

Conceptual Study

A person or group petitions NASA with an idea or plan. The proposal is studied and evaluated for merit, and, if accepted, the task of screening feasibility is delegated to a NASA Center. In the case of robotic deep space exploration, that center is frequently JPL. Sometimes it is APL, [1] or another center.

Prior to Phase A, the following activities typically take place: NASA Headquarters establishes a Science Working Group (SWG). The SWG develops the science goals and requirements, and prepares a preliminary scientific conception of the mission. Based on the high-level concept and the work of the SWG, a scientific document called the Announcement of Opportunity (AO)[2] is sent out by NASA Headquarters to individual scientists at universities, NASA centers, and science organizations around the world. The AO defines the existing concept of the mission and the scientific opportunities, goals, requirements, and system concepts. The AO specifies a fixed amount of time for the scientific community to respond to the announcement.

All proposals for new experiments are reviewed for science merit as related to the goal of the mission. Mass, power consumption, science return, safety, and ability to support the mission from the "home institution" are among key criteria. JPL develops a library of launch possibilities which becomes available to the project. Depending on the nature of the tasks at hand, they are delegated to various sections within JPL.

Historically, a project gets its start when funding is made available to JPL's Mission Design Section. The Mission Design Section then tasks personnel from appropriate divisions or sections as needed. For example, the Spacecraft Systems Engineering Section for Spacecraft Design, the Navigation Systems Section for Navigation Design, and the Mission Execution and Automation Section for Mission Operations. Note that the names of JPL sections change over the years to accommodate the Laboratory's ever-evolving management structure.

Usually the presentation of the study concept to NASA Headquarters by JPL personnel and NASA's approval to proceed to Phase A signify the end of Conceptual Study.

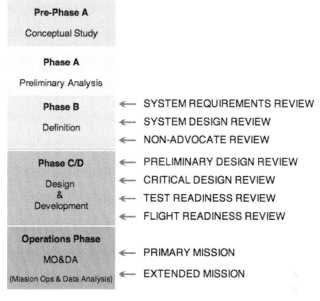

Figure 7.1: Full Mission Life Cycle.

Phase A: Preliminary Analysis

The Project creates a preliminary design and project plan as a proof of concept specifying what to build, when to launch, the course the spacecraft is to take, what is to be done during cruise, when the spacecraft will reach the target, and what operations will be carried out. The preliminary plan also addresses build-versus-buy decisions, what spacecraft instruments are needed, where system tests will be performed, who performs mission operations, what ground data system capabilities are required, and who the experimenters are. Generally speaking, publication of the preliminary plan with costing data marks the completion of Phase A: Preliminary Analysis.

Phase B: Definition

The definition phase converts the preliminary plan into a baseline technical solution. Requirements are defined, schedules are determined, and specifications are prepared to initiate system design and development. Major reviews commonly conducted as part of the definition phase are: System Requirements Review, System Design Review, and Non-Advocate Review.

The proposed experiments are divided into two classes based on facilities and experimenters. The facilities form teams around a designated set of hardware. Facilities are selected based on existing resources and past performance. Experimenters were specified in the preliminary plan. However, individuals are encouraged to respond with modifications and to step forward with their own ideas. These ideas could include the addition of another experiment.

A NASA peer group reviews all new proposals and "grades" them. After that, a sub-committee from NASA Headquarters' Science Mission Directorate (SMD) Steering Committee (SC) makes the final experiment selection, based on scientific value, cost, management, engineering, and safety.

Personnel teams are established to build and operate the instruments and evaluate the data returned. There is usually one team for each experiment, with one individual from that team chosen as the Team Leader (TL) and/or Principal Investigator (PI). In most cases, the Non-Advocate Review marks the end of Phase B: Definition.

Phase C/D: Design and Development

During the design and development phase, schedules are negotiated, and the space flight system is designed and developed. The phase begins with building and integrating subsystems and experiments into a single spacecraft. In a process called ATLO (Assembly, Test, and Launch Operations) the spacecraft is assembled integrated, tested, launched, deployed, and verified. Prior to launch, though, the complete spacecraft is tested together in a simulated interplanetary space environment. Voyager, Ulysses, Galileo, Cassini, and many more spacecraft have undergone extensive testing in JPL's 25-foot diameter solar-thermal-vacuum chamber. Figure 7.2 shows the Galileo spacecraft preparing for its turn.

Ground systems to support the mission are also developed in parallel with the spacecraft development, and are exercised along with the spacecraft during tests. Phase C/D typically lasts until 30 days after launch. Reviews commonly conducted as part of the design and development phase include: Preliminary Design Review, Critical Design Review, Test Readiness Review, and Flight Readiness Review.

Figure 7.2: Before being launched from the shuttle, the Galileo orbiter was tested in a space simulation chamber. The test chamber, located at JPL, is designed to subject spacecraft to approximately the same environmental conditions they will encounter in space. The high-gain antenna can be seen fully extended. The brightness and contrast of this image have been increased to expose detail of the spacecraft even though the high-gain antenna and other features remain overexposed in the bright simulated sunlight of the chamber.

Operations Phase

Operations phase E is also called MO&DA for Mission Operations and Data Analysis. It includes flying the spacecraft and obtaining science data for which the mission was designed. This phase is described in later sections of this training module: Chapters 14 through 17 present details of Launch, Cruise, Encounter, Extended Operations, and Project Closeout.

Design Considerations

The process by which a mission is conceived and brought through the phases described above includes consideration of many variables. The remainder of this chapter simply touches upon a few of them.

Budget

Trajectories are constrained by the laws of celestial mechanics, but the realities of budgets constrain the desires and needs of project science to determine the final choices. Should the mission use a quick, direct path that can be achieved only with a massive upper stage, or an extended cruise with gravity assists for "free" acceleration? Can significant science be accomplished by going only a few weeks out of the way? Which options can be justified against the cost in personnel and time? This task of balancing the monetary, the political and the physical is ordinarily resolved before most project personnel are assigned. The Cassini mission underwent a design change early in its life cycle, reducing its hardware complexity and saving substantial cost initially (although simplifying the hardware did increase the mission's later operational costs somewhat).

Design Changes

The purpose, scope, timing and probable budget for a mission must be clearly understood before realistic spacecraft design can be undertaken. But even a final, approved and funded design may be altered when assumed conditions change during its lifetime. Late design changes are always costly. The Galileo mission design, for example, underwent many significant and costly changes before it was finally launched. The Space Station Freedom spent tens of billions of dollars over several years prior to having comprehensive design changes imposed and becoming the International Space Station.

Resource Contention

Timing for many JPL missions is affected most directly by solar system geometry, which dictates optimum launch periods. It correspondingly implies the "part of the sky" that the proposed spacecraft will occupy and how many other spacecraft it may have to compete with for DSN antenna time. If possible, it is very advantageous to fly a mission toward an area where the spacecraft will share little or none of its viewperiod with other missions. Viewperiod is the span of time during which one DSN station can observe a particular spacecraft above its local horizon — perhaps eight to twelve hours each day.

Years before launch, mission designers request a "what-if" study by JPL's Resource Analysis Team to determine the probable degree of contention for DSN tracking time during the mission. Such a study can assist project management in the selection of launch date and mission profile with

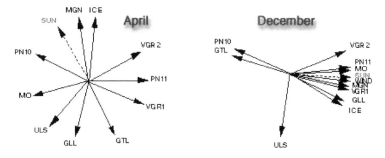

Figure 7.3: Spacecraft Right Ascension Chart

the least contention for external resources, and maximized science return for the mission.

Figure 7.3 illustrates how viewperiods may cause different spacecraft to compete for DSN resources. Abbreviations may be found in the Glossary. Even though the snapshot above dates back to 1993, it presents a valid picture of how various missions have to face potential contention for resources. When spacecraft occupy different areas of the sky, as in the April example, contention is at a minimum. However, when several spacecraft are bunched together in the same part of the sky, as they are in the December example, contention for DSN resources within heavily populated bunches may be formidable. Diagrams such as those shown above are produced by the Resource Allocation Team for ten year periods. They represent the situation on the 15th day of the month shown. The arrow indicates the center of a spacecraft view from Earth. Extend 60 degrees on both sides of an arrow to describe an 8-hour viewperiod for a spacecraft.

Tracking Capabilities

DSN tracking and data handling capabilities must be considered when designing on-board storage, telemetry rates, trajectory and launch periods. Magellan, for example, acquired radar data at 800 kilobits per second. Since it used its high-gain antenna for both mapping and high rate communications, it required on-board storage sufficient to record its data during each mapping pass. The project needed assurance that it could count on DSN tracking time nearly 24 hours a day for the duration of the mission. The data for each orbit had to be downlinked immediately after being acquired or it would be lost, overwritten by data from the next orbit. This scheme made good use of the highly elliptical orbit that Magellan occupied during mapping phase. High-rate data acquisition took place during the 20 or 30 minutes near periapsis, and the hour-long outbound and inbound legs of each orbit were necessary to transmit the data to Earth at the lower rate of 268.8 kbps.

The Mars Global Surveyor spacecraft also had limited on-board storage that required carefully planned DSN tracking frequency to avoid data loss. The high transmission rate and the maximum distance to Mars must be taken into account when designers determine such things as transmitter power and high-gain antenna size. Most planetary orbiters, including Galileo and Cassini, face similar tracking and data delivery constraints.

Data Return

The proposed volume and complexity of the mission's telemetry influences the cost of ground processing. If telemetry does not present significant differences from recent missions, it may be economical to use an adaptation of the existing Advanced Multimission Operations System (AMMOS) rather than develop one that is mission-specific. In 1985 the Infrared Astronomical Satellite (IRAS) mission's data requirements drove the implementation of an entire data processing facility on the Caltech campus. Known as the Infrared Processing and Analysis Center (IPAC), it is also being used to support the Space Infrared Telescope Facility (Spitzer) mission.

Notes

[1] http://www.jhuapl.edu
[2] http://heasarc.gsfc.nasa.gov/mail_archive/xmmnews/msg00063.html

Chapter 8

Experiments

> **Objectives:** Upon completion of this chapter you will be able to identify what is referred to as the scientific community, describe the typical background of principal investigators involved with space flight, and describe options for gathering science data. You will be aware of radio science's special capabilities, and you will be able to describe avenues for disseminating the results of science experiments.

Obtaining **scientific information** is the primary reason for launching and operating a robotic deep-space mission. Information is obtained by conducting experiments under controlled conditions to collect data. After the experimenters have had a chance to examine their data, and to perform extensive analysis, that information is made available to the larger science community for peer review, and generally at that time to the public at large. Sometimes, however, because of great public interest in a landing or encounter, JPL's Public Information Office, or other institutions, will release imaging and other data to the media shortly after it arrives on Earth, before investigators have had much chance to perform their analysis and invite peer review.

Figure 8.1: At JPL, Cornell professor Steve Squyres waits hopefully for Mars Rover egress.

The Scientific Community

The scientific community involved in JPL's experiments is worldwide and typically is composed of PhD-level scientific professionals tenured in academia, and their graduate students, as well as similar-level professional scientists and their staff from industry, scientific institutions, and professional societies. By way of illustration, here are a few random examples of facets of the world's scientific community, who might typically become involved in JPL robotic missions:

- Space Science Institute[1]

- Max-Planck Institut für Kernphysik[2]

- University of Iowa Physics[3]

- Southwest Research Institute[4]

- Caltech PMA (Physics, Math, Astronomy Division)[5]

- Malin Space Science Systems[6]

Gathering Scientific Data

Some experiments have a dedicated instrument aboard the spacecraft to measure a particular physical phenomenon, and some do not. A designated principal investigator (PI) or in many cases, a team, determines or negotiates the experiment's operation and decides who will analyze its data and publish the scientific results. Members of these teams may have been involved in the design of the instrument. Some examples of this kind of experiment are:

- The Radar Sensor on the Magellan spacecraft and the associated Radar Investigation Group of 26 scientists worldwide headed by a PI at MIT;

- The Low Energy Charged Particle experiment on the Voyager spacecraft and the associated PI at Johns Hopkins University;

- The Photopolarimeter experiment on the Voyager spacecraft and the PI at JPL;

- The Solid State Imaging experiment on the Galileo spacecraft and the imaging team headed by a PI at the University of Arizona.

Details of typical individual instruments aboard spacecraft that are used to gather data for these experiments appear in Chapter 12.

Other experiments are undertaken as opportunities arise to take advantage of a spacecraft's special capabilities or unique location or other circumstance. Some examples of this kind of experiment are:

- The gravitational wave search using the DSN and telecommunications transceivers aboard the Ulysses, Mars Observer, Galileo, and Cassini spacecraft (the PI is at Stanford University);

- The UV spectral observations of various astronomical sources using the Voyager UV spectrometer by various members of the astronomical community; and

- Venus atmospheric density studies using the attitude reaction wheels aboard the Magellan spacecraft by the PI at the NASA Langley Research Center.

- Solar corona studies conducted by a PI at JPL using DSN's radio link with various interplanetary spacecraft as they pass behind the Sun.

Science and Engineering Data

Data acquired by the spacecraft's scientific instruments and telemetered to Earth, or acquired by ground measurements of the spacecraft's radio signal in the case of Radio Science, in support of scientific experiments, is referred to as science data. Science data is the reason for flying a spacecraft.

The other category of data telemetered from a spacecraft, its health and status data such as temperatures, pressures, voltages, and computer states, is referred to as engineering data. This is normally of a more repetitive nature, and if some is lost, the same measurement of pressure or temperature can be seen again in a short time.

Except in cases of spacecraft anomalies or critical tests, science data is always given a higher priority than engineering data, because it is a mission's end product. Engineering data is used in carrying out spacecraft operations involved in obtaining the science data.

The Science Data Pipeline

Science data from on-board instruments, once received at the antennas of the DSN, flows through a string of computers and communications links known collectively as the Deep Space Mission System (DSMS), formerly called the

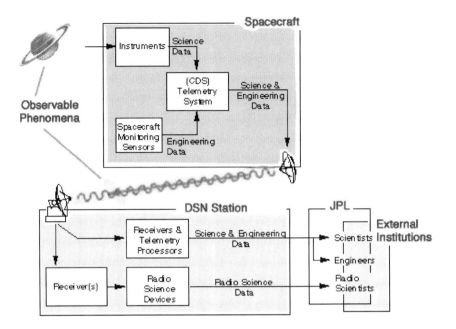

Ground Data System (GDS). The functions of the DSMS can be viewed as generally divided into two high-level segments: front-end and back-end.

Front-end processing consists of frame-synchronizing the data stream (discussed further in Chapter 18), restoring the data formats that were created by the spacecraft computers, and providing real-time visibility of engineering and tracking data for engineering analysts and science instrument teams. Back-end processing consists of data management to provide complete and catalogued data sets, production of data products such as images, and use of tools to access the data storage and cataloguing systems.

While there is typically some front-end visibility into some of the science data in real time, it is mainly through the back end systems that science teams (for whom the missions are flown) formally have access to complete sets of their science data.

Data Gaps

When science data is first received and stored by the data management system, it is common for some segments to be missing. A data management team or individual determines what gaps exist and whether or not they're recoverable.

- Data that is easily recoverable has reached the ground and was stored either at a DSN station or at some intermediate subsystem in the DSMS front end. The data may have been missing from the back end due to some failure in the pipeline. Once identified and located recovered data can be transferred to the data management system storage and integrated with data received earlier. This is typically a highly automated protocol or process.

- The problem is more complicated if DSN station problems or sudden rain over a station prevented reception of the data. In such cases, if it is of great value, the project may be able to recover it by commanding the spacecraft to replay a specific portion from its on-board storage subsystem before it gets overwritten.

Final science data products usually consist of time-ordered, gap-controlled sets of instrument-specific data records known as Experiment Data Records (EDRs). Other products that support analysis of the science data include collections of DSN monitor data which indicates the performance of DSN receivers, tracking and telemetry equipment, selected spacecraft engineering data, spacecraft ephemeris and pointing data. These are known as Supplementary Experiment Data Records (SEDRs) or the equivalent. SEDRS track the history of instrument pointing (discussed in Chapter 12), detailing the instrument's "footprint" on the object being imaged.

Spacecraft, Planet, Instruments, C-matrix, and Events (SPICE) kernels (files) provide spacecraft and planetary ephemerides, instrument mounting alignments, spacecraft orientation, spacecraft sequences of events, data needed for certain time conversions, etc. SPICE kernels are produced by the JPL Navigation and Ancillary Information Facility (NAIF) and are archived by the PDS NAIF Node. Related files include the Spacecraft and Planet ephemeris data Kernel (SpK), and the Planet Physical and Cartographic Constraints PcK Kernel.

Science data products have historically been produced within the Data Management Systems of flight projects. Cassini uses an example of a new, more distributed plan that calls for its diverse science teams to produce most or all the science data products after compilation and analysis. Cassini's small Data Management Team performs only those data management functions needed to deliver complete data sets to the science teams.

Radio Science

Radio science (RS) experiments use the spacecraft radio and the DSN together as their instrument, rather than using only an instrument aboard
the spacecraft. They record the attenuation, scintillation, refraction, rotation, Doppler shifts, and other direct modifications of the radio signal as
it is affected by the atmosphere of a planet, moons, or by structures such
as planetary rings or gravitational fields. From these data, radio scientists
are able to derive a great deal of information such as the structure and
composition of an atmosphere and particle sizes in rings.

Occultations

One RS experiment can take place when a
spacecraft passes behind a ring system. The
spacecraft keeps its radio signal trained on
Earth, and effects of ring particles can be detected in the signal. Passing behind a planet
or a planet's atmosphere, the spacecraft may
be commanded to rotate so that its radio signal remains trained on the "virtual Earth,"
the point where refraction makes the signal
bend toward Earth.

Figure 8.2: Radio Science Atmospheric Occultation Experiment.

Solar Corona

The "atmosphere" of the Sun is another target of great interest which can
be observed by Radio Science when a spacecraft is near superior conjunction. The solar corona causes scintillation, rotation, and other effects on the
spacecraft's radio signal which can be measured while a spacecraft is within
a few tens of degrees from the Sun as viewed from Earth. This data is useful
for investigating the nature and behavior of the solar corona.

Gravitational Lensing

Also when a spacecraft is near superior conjunction, Radio Science experiments may be conducted to quantify the general-relativistic gravitational
bending (also called gravitational lensing) imposed on the spacecraft's radio
link as it grazes the Sun. Such bending results in a slight increase in the
apparent distance to the spacecraft. This predicted GR effect was first con-

firmed in 1919 during a solar eclipse by observing the shift in the apparent positions of stars in the Hyades cluster visible near the eclipsed solar disc.

Gravitational Waves

Another RS experiment is the gravitational wave search, or gravitational wave experiment (GWE). Gravitational waves are predicted by general relativity, but as of early 2011 they have never been detected directly. Measuring minute Doppler shifts of a spin-stabilized spacecraft or a reaction-wheel-stabilized spacecraft in interplanetary space over long periods of time might yield the discovery. Spacecraft such as Voyager, which are stabilized by thruster bursts, cannot participate.

The spacecraft's distance would be observed to increase and then decrease on the order of millimeters as a gravitational wave passes through the solar system. Even if these gravitational wave searches have negative results, this information is scientifically useful because it places limits on the magnitude of long-wavelength gravitational waves. Dedicated observatories on Earth[7] and in space[8] are being readied to participate in the search for gravitational radiation.

Celestial Mechanics

Another RS experiment can determine the mass of an object like a planet or satellite. When nearing the body, the experiment measures the minute acceleration its gravitation exerts on the spacecraft. This acceleration is translated into a measurement of the body's mass. Once the mass is known, by the way, the object's density can be determined if images are available that show the body's size. Knowing the density provides powerful clues to a body's composition.

Gravity Field Surveys

Another science experiment, like radio science but not strictly classified as such, does not use an instrument aboard the spacecraft. Gravity field surveys (not to be confused with gravitational wave searches) use the spacecraft's radio and the DSN to measure minute Doppler shifts of a vehicle in planetary orbit. After subtracting out the Doppler shifts induced by planetary

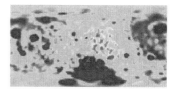

Figure 8.3: Crustal thickness on the Moon as derived from gravity field survey data.[9]

movement, the spacecraft's primary orbital motion, and small force factors
such as the solar wind and atmospheric friction, the residual Doppler shifts
are indicative of small spacecraft accelerations and decelerations. These
are evidence for variations in the planet's gravity field strength associated
with high and low concentrations of mass at and below the planet's surface.
Mapping the planet's mass distribution in this way yields information that
complements other data sets such as imaging or altimetry in the effort to
understand geologic structure and processes at work on the planet. Gravity
field surveying is further described in Chapter 16.

Dissemination of Results

Publication of the results of the experiments takes place in the literature
of the scientific community, notably the journals Science (American Association for the Advancement of Science, AAAS), Nature, the international
weekly journal of science, JGR (Journal of Geophysical Research, a publication of the American Geophysical Union), and Icarus, the official publication
of the of the American Astronomical Society.

Experimenters whose instruments ride aboard JPL spacecraft make pre-
sentations at virtually every periodic convention of the various scientific organizations, such as those mentioned above. If an operations person has
an opportunity to attend one or more, it would be very worthwhile. The
news media and several magazines keep a close eye on all of these journals
and proceedings and report items of discovery from them. The thin weekly
magazine Science News is a notable example, as is the amateur astronomers'
monthly Sky & Telescope magazine. Splendid photography from JPL's missions occasionally appears in National Geographic magazine, and many a
JPL mission has enjoyed very good treatment in public television's science
series Nova and documentaries on cable TV's Discovery channel.

Regional Planetary Imaging data facilities (RPIF) [10] are operated by
NASA's Planetary Data System (PDS) [11] at over a dozen sites around the
United States and overseas. Each maintains a complete photographic library
of images from NASA's lunar and planetary missions. They are open to
members of the public by appointment for browsing, and their staff can assist
individuals in selecting and ordering materials. All of NASA's planetary
imaging data is made available for researchers who are funded by NASA, in
photographic format and digital data format, via the PDS.

Educators may obtain a wide variety of materials and information from
NASA's flight projects through the network of Educator Resource Cen-

ters (ERC) in cooperation with educational institutions around the country. ERCs also support a center for distribution of multimedia resource materials for the classroom, called Central Operation of Resources for Educators (CORE). [12] JPL's Education website[13] offers more information about various avenues for information dissemination.

Notes

[1] http://www.spacescience.org
[2] http://www.mpi-hd.mpg.de
[3] http://www.physics.uiowa.edu
[4] http://www.swri.org
[5] http://www.pma.caltech.edu
[6] http://www.msss.com
[7] http://www.ligo.caltech.edu
[8] http://lisa.jpl.nasa.gov
[9] Image courtesy Clementine Mission, which was jointly sponsored by the Ballistic Missile Defense Organization and NASA.
[10] http://www.lpi.usra.edu/library/RPIF
[11] http://pds.jpl.nasa.gov
[12] http://www.nasa.gov/offices/education/programs/national/core
[13] http://www.jpl.nasa.gov/education

Chapter 9

Spacecraft Classification

> **Objectives:** Upon completion of this chapter you will be able to state the characteristics of various types of robotic spacecraft and be able to identify any of JPL's past, current, or future spacecraft as belonging to one of eight basic categories.

Robotic spacecraft are specially designed and constructed systems that can function in specific hostile environments. Their complexity and capabilities vary greatly and their purposes are diverse. To make some sense of all these variables, this chapter arbitrarily designates eight broad classes of robotic spacecraft according to the missions the spacecraft are intended to perform:

1. Flyby spacecraft
2. Orbiter spacecraft
3. Atmospheric spacecraft
4. Lander spacecraft
5. Penetrator spacecraft
6. Rover spacecraft
7. Observatory spacecraft
8. Communications & Navigation spacecraft

We illustrate these eight classes by offering one prime example of each, pictured on this page plus some additional examples in this chapter. The JPL public website has an up-to-date listing of all past, current, future and proposed JPL robotic spacecraft missions.[1] From there you can select and read more about each prime example, and additional missions. Spacecraft that carry human occupants are not considered here.

(1) Flyby Spacecraft

Flyby spacecraft conducted the initial reconnaissance phase of solar system exploration. They follow a continuous solar orbit or escape trajectory, never to be captured into a planetary orbit. They must have the capability of using their instruments to observe targets they pass. Ideally, their optical instruments can pan to compensate for the target's apparent motion in the instruments' field of view. They must downlink data to Earth, storing data onboard during the periods when their antennas are off Earthpoint. They must be able to survive long periods of interplanetary cruise. Flyby spacecraft may be designed to be stabilized in 3 axes using thrusters or reaction wheels, or to spin continuously for stabilization.

Figure 9.1: Voyager.

Our prime example of the flyby spacecraft category is the pair of Voyager spacecraft, which conducted encounters in the Jupiter, Saturn, Uranus, and Neptune systems. The Voyager 1 and 2 spacecraft are detailed on page 135. Other examples of flyby spacecraft include:

- Stardust Cometary Sample Return
- Mariner 2 to Venus
- Mariner 4 to Mars
- Mariner 5 to Venus
- Mariner 6 and 7 to Mars
- Mariner 10 to Mercury
- Pioneers 10 and 11 to Jupiter and Saturn
- New Horizons Pluto-Kuiper Belt Mission

(2) Orbiter Spacecraft

A spacecraft designed to travel to a distant planet and enter into orbit about it, must carry a substantial propulsive capability to decelerate it at the right moment, to achieve orbit insertion. It has to be designed to live with the fact that solar occultations will occur, wherein the planet shadows the spacecraft, cutting off any solar panels' production of electrical power and subjecting the vehicle to extreme thermal variation. Earth occultations will also occur, cutting off uplink and downlink communications with Earth. Orbiter spacecraft are carrying out the second phase of solar system exploration, following up the initial reconnaissance with in-depth study of each of the planets. The extensive list includes Magellan, Galileo, Mars Global Surveyor, Mars Odyssey, Cassini, and Messenger.

Our prime example of the orbiter spacecraft category is Galileo which entered orbit about Jupiter in 1995 to carry out a highly successful study of the Jovian system. Details of the Galileo spacecraft can be found on the project's website.[2] Other examples of orbiter spacecraft include:

Figure 9.2: Galileo.

- Messenger Mercury Orbiter
- Mariner 9 Mars Orbiter
- Cassini Saturn Orbiter
- Mars Global Surveyor
- Mars Odyssey
- TOPEX/Poseidon Earth Orbiter
- Ulysses Solar Polar Orbiter
- Jason Earth Orbiter
- 2001 Mars Odyssey
- Magellan Venus Orbiter
- Mars Observer a spacecraft (lost)

(3) Atmospheric Spacecraft

Atmospheric spacecraft are designed for a relatively short mission to collect data about the atmosphere of a planet or satellite. One typically has a limited complement of spacecraft subsystems. For example, an atmospheric spacecraft may have no need for propulsion subsystems or attitude and articulation control system subsystems at all. It does require an electric power supply, which may simply be batteries, and telecommunications equipment for tracking and data relay. Its scientific instruments may take direct measurements of an atmosphere's composition, temperature, pressure, density, cloud content and lightning.

Typically, atmospheric spacecraft are carried to their destination by another spacecraft. Galileo carried its atmospheric probe on an impact trajectory with Jupiter in 1995 and increased its spin rate to stabilize the probe's attitude for atmospheric entry. After probe release Galileo maneuvered to change from an impact trajectory to a Jupiter Orbit Insertion trajectory. An aeroshell protected the probe from the thousands of degrees of heat created by atmospheric compression during atmospheric entry, then parachutes deployed after the aeroshell was jettisoned. The probe completed its mission on battery power, and the orbiter relayed the data to Earth. The Pioneer 13

Venus Multiprobe Mission deployed four atmospheric probes that returned data directly to Earth during descent into the Venusian atmosphere in 1978.

Balloon packages are atmospheric probes designed for suspension from a buoyant gas bag to float and travel with the wind. The Soviet Vega 1 and Vega 2 missions to Comet Halley in 1986 deployed atmospheric balloons in Venus' atmosphere en route to the comet. DSN tracked the instrumented balloons to investigate winds in the Venusian atmosphere. (The Vega missions also deployed Venus landers.) While not currently funded, informal plans for other kinds of atmospheric spacecraft include battery powered instrumented airplanes and balloons for investigations in Mars' atmosphere.

Our prime example of the atmospheric spacecraft category is Huygens, which was carried to Saturn's moon Titan by the Cassini spacecraft. See page 136 for details of the Huygens spacecraft. Other examples of atmospheric spacecraft include:

Figure 9.3: Huygens.

- Galileo Atmospheric Probe
- Mars Balloon
- Titan "Aerover" Blimp
- Vega Venus Balloon
- JPL Planetary Aerovehicles Development
- Pioneer 13 Venus Multiprobe Mission

(4) Lander Spacecraft

Lander spacecraft are designed to reach the surface of a planet and survive long enough to telemeter data back to Earth. Examples have been the highly successful Soviet Venera landers which survived the harsh conditions on Venus while carrying out chemical composition analyses of the rocks and relaying color images, JPL's Viking landers at Mars, and the Surveyor series of landers at Earth's moon, which carried out similar experiments. The Mars Pathfinder project, which landed on Mars on July 4, 1997, was intended to be the first in a series of landers on the surface of Mars at widely distributed locations to study the planet's atmosphere, interior, and soil. The lander, carrying its own instruments, was later named the Carl Sagan Memorial Mars Station. Pathfinder also deployed a rover, Sojourner. A system of actively-cooled, long-lived Venus landers designed for seismology investigations, is being envisioned for a possible future mission.

Our prime example of the lander spacecraft category is Mars Pathfinder. See page 138 for details of the Pathfinder spacecraft. Other examples of lander spacecraft include:

- Viking Mars Landers
- Venera 13 Venus Lander
- Surveyor Moon Landers

Figure 9.4: Pathfinder.

(5) Penetrator Spacecraft

Surface penetrators have been designed for entering the surface of a body, such as a comet, surviving an impact of hundreds of Gs, measuring, and telemetering the properties of the penetrated surface. As of April 2006, no Penetrator spacecraft have been successfully operated. Penetrator data would typically be telemetered to an orbiter craft for re-transmission to Earth. The Comet Rendezvous / Asteroid Flyby (CRAF) mission included a cometary penetrator, but the mission was cancelled in 1992 due to budget constraints.

Our prime example of a penetrator spacecraft is the twin Deep Space 2 penetrators which piggybacked to Mars aboard the Mars Polar Lander and were to slam into Martian soil December 3, 1999. They were never heard from. See page 139 for details of the penetrator spacecraft. Other examples of penetrator spacecraft include:

- Deep Impact Mission to a comet
- Ice Pick Mission to Europa
- Lunar-A Mission to Earth's Moon

Figure 9.5: DS-2.

(6) Rover Spacecraft

Electrically-powered rover spacecraft are being designed and tested by JPL as part of the Mars exploration effort. The Mars Pathfinder project included a small, highly successful mobile system referred to as a micro-rover by the name of Sojourner. Mars rovers are also being developed by Russia with a measure of support from The Planetary Society. Rover craft need to be be semi-autonomous. While they are steerable from Earth, the delay inherent in radio communications between Earth and Mars means they must be able to make at least some decisions on their own as they move. Their purposes range from taking images and soil analyses to collecting samples for return to Earth.

Our prime example of a rover spacecraft is of course the famous Sojourner Rover, shown here in an image from the surface of Mars. See page 140 for details of this rover spacecraft. Other examples of rover spacecraft include:

- Mars Exploration Rovers Spirit, Opportunity
- Lunokhod and Marsokhod Russian Rovers
- JPL Inflatable Rovers
- Red Rover Student activity

Figure 9.6: Sojourner.

(7) Observatory Spacecraft

An observatory spacecraft does not travel to a destination to explore it. Instead, it occupies an Earth orbit, or a solar orbit, from where it can observe distant targets free of the obscuring and blurring effects of Earth's atmosphere.

NASA's Great Observatories program studies the universe at wavelengths from infra-red to gamma-rays. The program includes four Observatory Spacecraft: the familiar Hubble Space Telescope (HST), the Chandra X-Ray Observatory (CXO, previously known as AXAF), the Compton Gamma Ray Observatory (GRO), and the Space Infrared Telescope Facility (SIRTF) renamed Spitzer in flight.

The HST is still operating as of early 2011. GRO has completed its mission and was de-orbited in June 2000. CXO was launched in July 1999 and

Figure 9.7: Spitzer.

continues to operate. SIRTF launched in January 2003 and is currently operating. In the coming decades many new kinds of observatory spacecraft will be deployed to take advantage of the tremendous gains available from operating in space.

Our prime example of an observatory spacecraft is the Spitzer Space Telescope. See page 141 for details of the observatory spacecraft. Other examples of observatory spacecraft include:

- HST Hubble Space Telescope
- Chandra X-ray Observatory
- Compton Gamma-ray Observatory
- IRAS Infrared Astronomical Satellite
- TPF Terrestrial Planet Finder
- SIM Space Interferometry Mission
- Planck Cosmic Background Radiation Field Survey
- NGST Next-Generation Space Telescope, renamed the James Webb Space Telescope, JWST

(8) Communications & Navigation Spacecraft

Communications and navigation spacecraft are abundant in Earth orbit, but they are largely incidental to JPL's missions. The Deep Space Network's ground communications systems do make use of Earth-orbiting communications spacecraft to transfer data among its sites in Spain, Australia, the California desert, and JPL. The Deep Space Network uses Earth-orbiting Global Positioning System navigation spacecraft to maintain an accurate time reference throughout the network.

Figure 9.8: TDRSS.

In the future, communications and navigation spacecraft may be deployed at Mars, Venus, or other planets, dedicated to communications with orbiters, rovers, penetrators, and atmospheric spacecraft operating in their vicinity. This task is currently carried out to some extent by various orbiter spacecraft that are also equipped for limited communications relay. The purpose of dedicated Mars communications orbiters would be to augment the Deep Space Network's capabilities to communicate with the resident spacecraft. None are have been funded or developed as of 2011. This concept is revisited in Chapter 18.

The communications spacecraft example offered here is NASA's Tracking and Data Relay Satellite System, TDRSS. NASA missions supported by the system include the Hubble Telescope, the Space Shuttle, GRO, Landsat,

TOPEX, and EUVE and the International Space Station. Details of this communications spacecraft may be found on the Goddard Space Flight Center website.[3] Other examples of communications and navigation spacecraft include:

- Milstar
- Global Positioning System (GPS)
- DirecTV
- Globalstar
- Iridium

For Further Reference

The Goddard Space Flight Center maintains a list of virtually every lunar and planetary mission ever flown or attempted by any nation, and those on schedule for future launch.[4] The list is arranged by launch date, and each entry is linked to a page of facts about the mission.

Voyagers 1 and 2

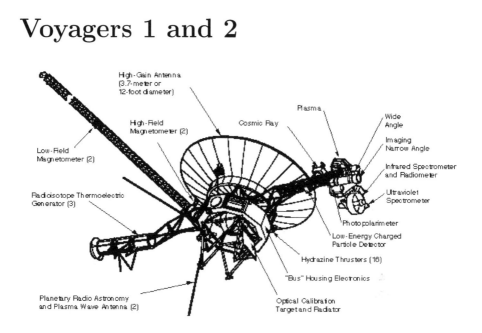

Classification: Flyby spacecraft.

Mission: Jovian planets and interstellar space.

Features: The Voyager 1 and Voyager 2 spacecraft were launched in late 1977 aboard Titan III launch vehicles with Centaur upper stages. They completed highly successful prime mission flybys of Jupiter in 1979 and Saturn in 1980 and 1981. Voyager 2's extended mission succeeded with flybys of Uranus in 1986 and Neptune in 1989. Both spacecraft are still healthy in 2011, and are conducting studies of interplanetary space enroute to interstellar space. Voyager 1 and Voyager 2 are collecting low frequency radio emissions from the termination shock (see Chapter 1), which is estimated to be within reach of the spacecraft during its mission lifetime. Science data return is expected to continue until about 2020.

A long-silent Voyager 2 is projected to pass within a light year of nearby Barnard's Star 350,000 years in the future. Both Voyager 1 and 2 will forever orbit the center of our galaxy, never returning to the Sun.

Stabilization: Three-axis stabilized via thrusters.

Project Website: http://voyager.jpl.nasa.gov

Huygens

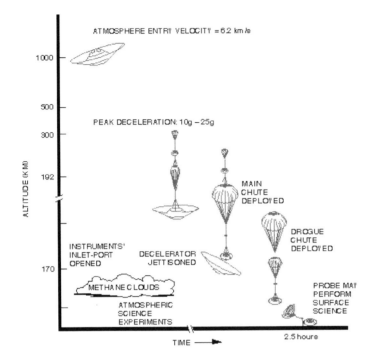

Classification: Atmospheric probe spacecraft.

Mission: Investigate Titan's atmosphere.

Features: The Huygens Probe, contributed by the European Space Agency
(ESA), was carried by the Cassini spacecraft to Titan, Saturn's largest moon,
and was deployed carrying six science instruments into Titan's atmosphere.
It returned a wealth of data to the Cassini Spacecraft which then relayed it
to Earth. It was released from Cassini on December 25 UTC, and carried
out its 2.5-hour mission, descending through and measuring Titan's atmosphere, on January 14, 2005. It also imaged the surface during descent.
Since Huygens actually survive impact with the surface of Titan, it was able
to continue to transmit science data from the surface for over 90 minutes.
Radio telescopes on Earth also received Huygens' tiny S-band signal directly
from Titan, providing data on Titan's winds via induced Doppler shifts.

Stabilization: Spin stabilized.

Project Website: http://www.esa.int/SPECIALS/Cassini-Huygens

Galileo Atmospheric Probe

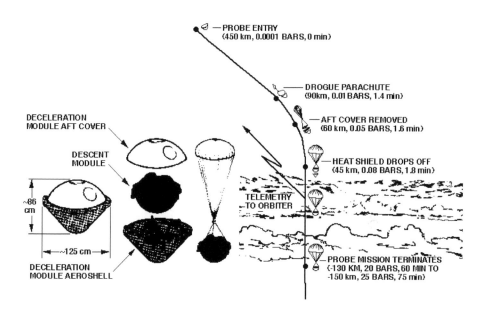

Classification: Atmospheric probe spacecraft.

Mission: Investigate Jupiter's upper atmosphere.

Features: The Galileo Atmospheric Probe was released from the Galileo orbiter spacecraft in July 1995, about 100 days before arrival at Jupiter. Atmospheric entry took place on 7 December, 1995, as the orbiter tracked the probe and recorded its data for later relay to Earth. Probe instruments investigated the chemical composition and the physical state of the atmosphere. The probe returned data for just under an hour before it was overcome by the pressure of Jupiter's atmosphere.

Stabilization: Spin stabilized.

Science Results Website: http://www.jpl.nasa.gov/sl9/gll38.html

Mars Pathfinder

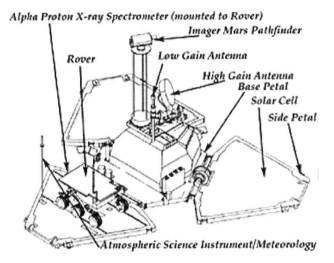

Classification: Lander spacecraft with surface rover.

Mission: Analyze Martian soil.

Features: Pathfinder was a low-cost mission with a single flight system launched December 1996 aboard a Delta rocket for a Mars landing in July 1997. The spacecraft entered the atmosphere directly from its transfer trajectory, and analyzed the atmosphere on the way in. It carried a small rover, which performed technology, science, and engineering experiments on the surface of Mars.

The lander parachuted toward the surface, with retrorocket braking assist. Eight seconds before impact on the Martian surface, three airbags inflated on each of the three folded "petals" of the lander, cushioning its impact. After the airbags deflated, the petals then deployed, exposing solar panels to the sunlight, and righting the lander. The rover then drove off the solar panel and onto the Martian soil. The lander was designed to operate on the surface for over 30 Martian days and nights, but it survived longer, returning panoramic views of the Martian landscape, and measuring the soil's chemistry, and characterizing the seismic environment.

Stabilization: Spin stabilized during cruise.

Project Website: http://mars.jpl.nasa.gov/MPF

Deep Space 2

Classification: Penetrator spacecraft.

Mission: Analyze Martian subsurface environment.

Features: Micro-instruments in the lower spacecraft (forebody) collect a sample of soil and analyze it for water content. Data from the forebody is sent through a flexible cable to the upper spacecraft (aftbody) at the surface; a telecommunications system on the aftbody relays data to the Mars Global Surveyor Spacecraft, operating in orbit on its mapping mission.

The Deep Space 2 probes were designed to use only an aeroshell. By eliminating parachutes and rockets, the probes are lighter and less expensive, but also very hardy. Similar in weight to a lap-top computer, they were designed to survive a high-speed impact, and to operate successfully in extremely low temperatures, something conventional miniaturized electronics, and standard spacecraft, could never do.

The two miniature probes, carrying ten experimental technologies each, piggybacked to the red planet with the Mars Polar Lander. They were to slam into Martian soil December 3, 1999. No communication was ever received from them, nor from the Mars Polar Lander.

Stabilization: Spin stabilized during cruise.

Project Website: http://nmp.jpl.nasa.gov/ds2

Sojourner

Classification: Rover spacecraft.

Mission: Analyze Martian soil and individual rocks.

Features: Sojourner weighs 11.0 kg on Earth and is about the size of a child's small wagon. It has six wheels and can move at speeds up to 0.6 meters per minute. The rover's wheels and suspension use a rocker-bogie system that is unique in that it does not use springs. Rather, its joints rotate and conform to the contour of the ground, providing the greatest degree of stability for traversing rocky, uneven surfaces.

Sojourner performed a number of science experiments to evaluate its performance as a guide to the design of future rovers. These included conducting a series of experiments that validate technologies for an autonomous mobile vehicle; deploying an Alpha Proton X-ray Spectrometer on rocks and soil; and imaging the lander as part of an engineering assessment after landing.

Also, Sojourner performed a number of specific technology experiments as a guide to hardware and software design for future rovers as well as assisting in verifying engineering capabilities for Mars rovers.

Stabilization: Spin stabilized during cruise.

Website: http://mars.jpl.nasa.gov/MPF/rover/sojourner.html

Spitzer Space Infrared Telescope

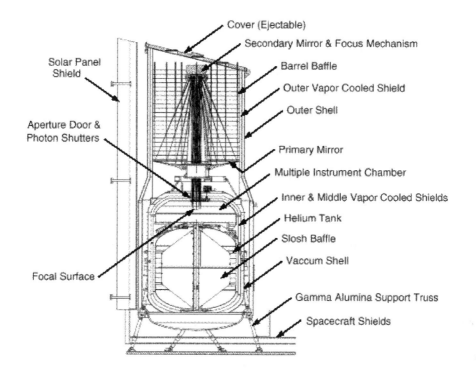

Classification: Observatory spacecraft.

Mission: Observe a variety of targets in the infrared.

Features: The Space InfraRed Telescope Facility (SIRTF) is a 950-kg space-borne, cryogenically-cooled infrared observatory capable of studying objects ranging from our Solar System to the distant reaches of the Universe. SIRTF is the final element in NASA's Great Observatories Program and an important scientific and technical cornerstone of the new Astronomical Search for Origins Program. SIRTF launched August 25, 2003. It occupies an Earth-trailing, heliocentric orbit. Its infrared instrument is cooled by 360 liters of liquid helium.

Stabilization: 3-axis stabilized using reaction wheels.

Project Website: http://www.spitzer.caltech.edu

Notes

[1]http://www.jpl.nasa.gov/missions
[2]http://solarsystem.nasa.gov/galileo
[3]http://nssdc.gsfc.nasa.gov/multi/tdrs.html
[4]http://nssdc.gsfc.nasa.gov/planetary/chronology.html

Chapter 10

Telecommunications

> **Objectives:** Upon completion of this chapter you will be aware of the major factors involved in communicating across interplanetary distances, including uplink, downlink, coherence, modulation, coding, and multiplexing. You will also be aware that additional coverage of this subject appears in different chapters.

T**his chapter offers a broad view** into some basic telecommunications issues, including both spacecraft and Earth-based components. Chapter 18 builds on the Deep Space Network's role in telecommunications, and more details of onboard spacecraft telecommunications equipment appear under Telecommunications Subsystems in Chapter 11.

Transmitters and Receivers

Chapter 6 showed that electromagnetic waves propagate any time there is a change in an electric current. Also, electromagnetic waves induce electric currents in conductors they encounter.

This is the basis for radio communications. A radio transmitter then, creates a changing electric current, and lets the resulting waves propagate. A radio receiver picks up the electric current induced by electromagnetic waves.

Since the wires in a house carry electric current that changes typically at the rate of 60 Hz (in the U.S., 50 Hz in some other locations), it is easy to pick up a small induced 60 Hz current in any free conductor anywhere in the house. Try it: a sensitive earphone or an amplifier reveals an audible 60-Hz hum, unless its input is carefully shielded.

Practical radio transmitters and receivers operate on the same basis, but at much higher frequencies than 60 Hz. With frequencies in the microwave region of the spectrum (see page 95) and higher, electromagnetic energy can easily be confined to a narrow beam, like a beam of light.

Signal Power

The KPFK FM radio broadcast transmitter on Mt. Wilson near Los Angeles has a power of 112 kW and a frequency of 90.7 MHz.[1] The transmitter is no more than 15 km away from JPL. Your automobile radio receiver has a simple antenna to pick up the induced signal.

By comparison, a spacecraft's transmitter might have no more than 20 W of radiating power, but it can bridge distances measured in tens of billions of kilometers. How can that be? One part of the solution is to employ microwave frequencies and use reflectors to concentrate all available power into a narrow beam, instead of broadcasting it in all directions. A spacecraft typically does this using a Cassegrain dish antenna, perhaps a few meters in diameter, trained precisely toward Earth.

The use of a cassegrain high-gain antenna (HGA) can increase telecommunications performance substantially. Consider a radio transmitter that radiates from a single point evenly in all directions. This theoretical concept (it cannot actually exist) is called an isotropic radiator. Voyager's, HGA, for example, provides a gain of 47 dBi at X-band. The reference "i" in dBi means the ratio is referenced to an isotropic radiator. The value of 47 dB (decibels) is $10^{47/10}$ is a factor of more than 50,000. A 20-Watt transmitter using a 47-dBi gain HGA would have an effective power of a megawatt along the highly directional beam. This effective power is called ite Equivalent Isotropically Radiated Power (EIRP).

Another component to the success of interplanetary communications is the fact that there are no significant sources of noise in interplanetary space at the spacecraft's specific frequency. The spacecraft is the only thing that can be detected at that frequency (unless it is happens to be close to the noisy Sun as seen in the sky).

The remainder of the solution is provided by the Deep Space Network's large aperture Cassegrain reflectors (the largest of which offer 74 dBi gain at X-band), cryogenically-cooled low-noise amplifiers, sophisticated receivers, and data coding and error-correction schemes. These systems can collect, detect, lock onto, and amplify a vanishingly small signal that reaches Earth from the spacecraft, and can extract data from the signal virtually without errors.

Uplink and Downlink

The radio signal transmitted from Earth to a spacecraft is known as uplink. The transmission from spacecraft to Earth is downlink. Uplink or downlink may consist of a pure RF tone, called a carrier. Such a pure carrier is useful in many ways, including radio science experiments. On the other hand, carriers may be modulated to carry information in each direction (modulation is detailed later in this chapter).

Commands may be transmitted to a spacecraft by modulating the uplink carrier.

Telemetry containing science and engineering data may be transmitted to Earth by modulating the downlink carrier.

Phase Lock

When your FM receiver tunes in a broadcast station, it locks onto the signal using an internal circuit called a phase-locked loop, PLL. (The PLL combines a voltage-controlled oscillator and a phase comparator designed so the oscillator tracks the phase of an incoming signal.) PLLs are used in spacecraft and DSN telecommunications receivers, and so it is common to speak of a spacecraft's receiver "locked" onto an uplink, and a DSN receiver "locked" onto a downlink. (Telemetry systems are said to achieve "lock" on data, as described later, but the concept is completely different from a receiver's phase-locked loop.)

One-way, Two-way, Three-way

When you're only receiving a downlink from a spacecraft, the communication is called one-way. When you're sending an uplink that the spacecraft is receiving at the same time a downlink is being received at Earth, the communications mode is called two-way.

There are finer points. The communications mode is still called one-way even when an uplink is being received by the spacecraft, but the full round-trip light time hasn't yet elapsed. Picture it this way: You're getting downlink and watching telemetry that shows the state of the spacecraft's own receiver. As long as you see that the spacecraft's receiver is not receiving the uplink, you're one-way. Once you see that the spacecraft's receiver has locked onto the uplink, you're two-way.

Three-way is when you're receiving a downlink on one station, but a different station is providing the uplink. Again, RTLT must have elapsed since the other station's uplink began. If you're watching telemetry from the spacecraft coming in over a DSN station in Australia, and you see the spacecraft's receiver is still in lock on the uplink provided by a DSN station at Goldstone, California, you're three-way.

Coherence

Aside from the information modulated on the downlink as telemetry, the carrier itself is used for tracking and navigating the spacecraft, as well as for carrying out some types of science experiments such as radio science or gravity field mapping. For each of these uses, an extremely stable downlink frequency is required, so that Doppler shifts on the order of fractions of a Hertz may be detected out of many GHz, over periods of many hours. But it would be impossible for any spacecraft to carry the massive equipment required to maintain such frequency stability. Spacecraft transmitters are subject to wide temperature changes, which cause their output frequency to drift.

Figure 10.1: DSN Hydrogen Maser

The solution is to have the spacecraft generate a downlink that is coherent to the uplink it receives.

Down in the basement of each DSN complex, there looms a hydrogen-maser-based frequency standard in an environmentally controlled room, sustained by an uninterruptable power supply. This maser is used as a reference for generating an extremely stable uplink frequency for the spacecraft to use, in turn, to generate its coherent downlink. (It also supplies a signal to the master clock that counts cycles and distributes UTC time.) Its stability is equivalent to the gain or loss of 1 second in 30 million years.

Once the spacecraft receives the stable uplink frequency, it multiplies that frequency by a predetermined constant, and uses that value to generate its downlink frequency. This way, the downlink enjoys all the extraordinarily high stability in frequency that belongs to the massive, sensitive equipment that generated the uplink. It can thus be used for precisely tracking the spacecraft and for carrying out precision science experiments.

The reason a spacecraft's transponder multiplies the received frequency by a constant, is so the downlink it generates won't cause interference with the uplink being received. Cassini's X-band transponder, for example, uses the constant 1.1748999.

The spacecraft also carries a low-mass oscillator to use as a reference in generating its downlink for periods when an uplink is not available, but it is not highly stable, since its output frequency is affected by temperature variations. Some spacecraft carry an Ultra-Stable Oscillator (USO), discussed further in Chapter 16. But even the USO has nowhere near the ideal stability of a coherent link.

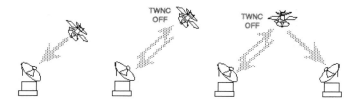

Figure 10.2: 1-Way, 2-Way Coherent, 2-Way and 3-Way Coherent.

Because of the stringent frequency requirements for spacecraft operations, JPL stays at the forefront of frequency and timing standards technology. Advances are being undertaken that may replace the hydrogen-maser-based system early in the 21st century.

Data Glitch Going Two-way

Consider a DSN station in lock with a spacecraft's one-way downlink. Now have that station send an uplink. When the spacecraft locks onto the uplink, it abandons the internal frequency reference it was using to generate its downlink. Now it uses the uplink to generate a new downlink frequency. That new downlink frequency will be a lot more stable, but it will likely be a slightly different frequency than the one it was generating on its own.

So when the new coherent downlink reaches the DSN station a round-trip light time after the DSN's transmitter goes on, the station's receiver drops lock on the old downlink, because it isn't there at the same frequency anymore. The DSN receiver has to change frequency, and lock on the new and different downlink frequency. Of course when the DSN receiver goes out of lock, the telemetry system also loses lock and the data stream stops.

It's as if you're tuned in to the KPFK FM radio broadcast at 90.700 MHz, and suddenly it decides to change to 90.750 MHz. You have to tune your radio to the new frequency, so you're going to miss some of that fascinating Pacifica Radio interview you were following!

The DSN station knows the exact time a coherent downlink will arrive, and will waste no time looking for the new frequency. It may be common, however, to experience a minute with no data while the lockup proceeds, so it is wise to plan for this outage in the early stages of determining what the content of the downlink will be. You don't want to sequence your spacecraft to be downlinking something important at the time it changes to 2-way coherent.

In order to avoid this data glitch when going from two-way coherent to three-way coherent, the two DSN stations coördinate closely to provide an uplink transfer. Picture a spacecraft setting at the western horizon as seen from the Goldstone DSN station. At the same time, the spacecraft is rising on the eastern horizon as seen from the Australian station. At a predetermined time, the Australian station turns on its transmitter, already tuned so it appears to the spacecraft as the same frequency as the uplink it is already receiving (taking into consideration the Doppler shifts induced as the turning Earth moves Australia towards, and Goldstone away from the spacecraft). Then, two seconds later, the Goldstone station turns off its transmitter. Nominally, the spacecraft will not lose lock on the uplink, and so its coherent downlink will not undergo any change in frequency. Telemetry and tracking data continue uninterrupted!

TWNC On

Most interplanetary spacecraft may also invoke a mode that does not use the uplink frequency as a reference for generating downlink (even if an uplink is present). When invoked, the spacecraft instead uses its onboard oscillator as a reference for generating its downlink frequency. This mode is known as Two-Way Non-Coherent (TWNC, pronounced "twink"). TWNC's default state is typically off, to enjoy the advantages of a coherent link.

When TWNC is on, the downlink is always non-coherent, even if the spacecraft sees an uplink.

Why have a two-way non-coherent capability? Mostly for short-duration Radio Science (RS) experiments. While RS usually needs the most stable coherent downlink available (referenced to extremely stable ground equipment with TWNC off), they can't always have one. Atmospheric occultation experiments provide an example. When a spacecraft goes behind a planet's atmosphere, the atmosphere corrupts the uplink before it reaches the spacecraft, adding to noise in the downlink. The solution is to turn coherency off (TWNC on, a double-negative phrase), and reference a known and well calibrated spacecraft-generated downlink, as a second-best source. Regularly scheduled non-coherent periods during cruise provide calibration data in support of non-coherent RS experiments.

Recall "two-way" and "three-way" mean there is an uplink and there is a downlink. These terms *do not* indicate whether the spacecraft's downlink is coherent to a station's uplink.

In common usage, some operations people often say "two-way" to mean "coherent." This is unfortunate. It can lead to confusion. Correctly stated, a spacecraft's downlink is coherent when it is two-way or three-way with TWNC off. With TWNC on, you can be two-way or three-way and non-coherent.

Modulation and Demodulation, Carrier and Subcarrier

To modulate means to modify or temper in some way. For example, consider a pure-tone carrier of, say, 3 GHz. If you were to quickly turn this tone off and on at the rate of a thousand times a second, we could say you are modulating the carrier with a frequency of 1 kHz using on-off keying.

Spacecraft and DSN carrier signals are modulated not by on-off keying as in the above example, but by shifting the waveform's phase slightly at a carefully clocked rate. This is phase modulation.[2] You can phase-modulate the carrier directly with data. Another scheme is to phase-modulate the carrier with a constant frequency called a subcarrier which in turn carries the data symbols.

In the on-off keying example, we could really call the 1-kHz modulation a subcarrier, in an attempt to simplify the explanation. Now, to stretch the example a bit, you might decide to place data onto this 1-kHz "subcarrier" by varying the timing of some of the on-off transitions. The same kind of schemes that put data on the downlink are also used on the uplink to carry commands or ranging tones for navigation. Ranging and other navigation topics appear in Chapter 13.

To illustrate the subcarrier scheme, let's say Cassini's X-band downlink carrier is 8.4 GHz. That's pretty close. Cassini's transmitter can impose a modulation of 360 kHz onto that carrier to create a subcarrier. Then, the data can be modulated onto the 360 kHz subcarrier by shifting its phase. In practice, Cassini can modulate data directly onto its carrier, or put it into either a 360 kHz subcarrier or a 22.5 kHz subcarrier. Which scheme is selected depends on the data rates to be used and other factors.

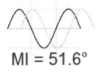

MI = 51.6°

The amount of phase shift used to modulate a carrier (or a subcarrier) is called modulation index (or mod index, MI). MI is measured in degrees or radians (see the illustration on page 104). The greater the mod index, the "louder" the modulation. Various MI values are used for different conditions. Demodulation is the reverse. It's the process of capturing data symbols from the carrier or subcarrier. It involves detecting the individual phase shifts in the carrier and subcarrier if present, and decoding them into digital data for further processing. The same processes of modulation and demodulation are used commonly with Earth-based computer systems and fax machines transmitting data back and forth over a telephone line. They use a modem, short for modulator / demodulator. Computer and fax modems modulate that familiar audio frequency carrier (you've heard it, it sounds like "beeeeeee, diddle-diddle-diddle") because the telephone system can readily handle it.

Modulation can be seen as power sharing. If there is no modulation, all the power is in the carrier. As you increase modulation, you put more power in the data and less in the carrier but it always adds up to the same amount of power. You decrease carrier power when you add modulation for telemetry, and decrease it again when you add modulation for ranging. When operating near the limits of reception, it may be necessary to choose between telemetry modulation and ranging modulation to preserve a useable downlink, and to choose between ranging modulation and command modulation to preserve a useable uplink.

Beacons

As part of its new technology, the Deep Space 1 spacecraft[3] has demonstrated beacon monitor operations, a mode which may become more widespread as spacecraft intelligence and capabilities increase.

In beacon monitor operations, an on-board data summarization system determines the overall spacecraft health. Then it selects one of 4 subcarrier tones to place on its downlink carrier to indicate whether, or how urgently, it needs contact using the Deep Space Network's larger antennas. These subcarrier tones are quickly and easily detected with low cost receivers and small antennas, so monitoring a spacecraft that uses this technology can free up precious resources of the Deep Space Network. Figure 10.3 shows the appearance of Deep Space 1's beacon.

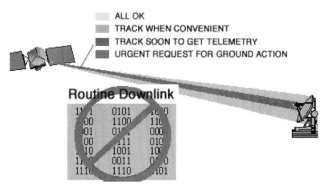

Each beacon tone is like a single note on a musical instrument. One tone might mean that the spacecraft is fine, and it does not need contact with human operators. Another might mean that contact is needed sometime within a month, while a third could mean that contact should be established within a week so that data can be downlinked. Another tone could mean "red alert," indicating the spacecraft needs immediate contact because of some problem.

Symbols and Bits and Coding

The Space Place has a good introductory lesson on bits and binary notation.[4]

Generally, modern spacecraft don't place data bits, as such, onto the carrier or subcarrier. They put symbols there instead. A symbol is a wiggle (or a non-wiggle) in the phase of the carrier or subcarrier. A number of

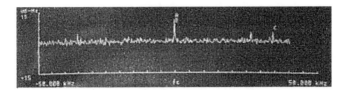

Figure 10.3: This is a photo of the downlink from the Deep Space 1 spacecraft displayed by an SSI (spectral signal indicator). This kind of display can be obtained quickly when the Deep Space Network antenna points to a spacecraft, since it does not require acquisition and processing of telemetry data. This makes it possible to use short periods of DSN tracking time to simply "look in" on the spacecraft's state of health.

The horizontal wiggly line represents mostly radio noise. The peak near the cursor labelled "AB" is the spacecraft's carrier at fc (X-band center carrier frequency) near the middle of the display. A smaller peak can be seen near the cursor labelled 'C' on the right. This peak near 40 kHz above fc and its counterpart on the left toward the "–50.000 kHz" label, constitutes a view of the beacon subcarrier. These labelled peaks can be seen to persist over time, while most of the smaller noise-peaks appear to change constantly. Simply viewing this display is enough to recognize which of a few different "beacon" signatures is present, therefore deducing the state of the spacecraft without having to lock up, process, and distribute telemetry data.

symbols make up a bit. How many depends on the coding scheme being used. Coding is a theoretical field using logic and mathematics to help ensure error-free data transmission. One coding scheme most interplanetary spacecraft use is a forward error correction (FEC) scheme called convolutional coding with Viterbi decoding.[5] Another kind is Reed-Solomon, the same coding your CD player uses to ensure error-free music. Reed-Solomon coding adds bits, but it doesn't affect the ratio of symbols to bits; it's generally imposed prior to the convolutional code. The flavor of convolutional coding that Cassini often uses (k=15, r=1/6) places six symbols on the downlink (carrier or subcarrier) for every bit of data. Thus Cassini's downlink data rate of 82,950 bps requires a symbol rate of 497,700 symbols per second. A recent advance in coding called turbo code is being used on some new missions.[6]

About those phase wigglings— Two schemes are commonly in use for putting them on the downlink or uplink:

Bi-phase, or bi-ϕ, is the format normally used for putting data directly on a carrier. In this scheme, the phase has to shift from one offset to another, across the zero point, to represent a symbol. The timing of the shift in relation to a symbol time (clock) determines whether the symbol is a "1" or a "0." This scheme is also called Manchester coding.

NRZ or non-return-to-zero, is the format normally used for putting data onto a subcarrier. In this scheme, a phase deviation is held for the duration of a symbol time (clock) to represent a logical "1." To represent a logical "0", the opposite phase deviation is held for a symbol time. If the symbols are "1110," the first shift is held for three symbol times without returning to zero.

In either case, a number of symbols must be received before a data bit can be recognized, the number depending on what convolutional coding parameters are in use.

Multiplexing

The subjects that follow are more properly described as issues of telemetry, rather than issues of basic telecommunications, which have been the focus until now.

Not every subsystem aboard a spacecraft can transmit its data at the same time, so the data is multiplexed. Cassini has over 13,000 different measurements such as pressures, temperatures, computer states, microswitch positions, science data streams, and so on. To ensure that all necessary data is received, and that important measurements are received more frequently than less important ones, spacecraft are designed to use one of two different multiplexing schemes, described below. Both methods make use of a data structure called a frame. The spacecraft puts a known bit pattern into the data stream to separate the stream into frames. This bit pattern is called a PN (Pseudo-Noise) code, or a sync marker. The ground system searches for this pattern, and identifies all data in between instances of this pattern as a frame. The two multiplexing methods differ in how the spacecraft organizes the bits within the frames, between the sync markers:

In TDM, or Time Division Multiplexed data, each measurement is commutated (picked up and selectively placed) into a known location in the frame. In a simplified example, it might be agreed that right after the PN code is sent, the imaging system sends 1024 bits of its data, then the propulsion system sends 512 bits, and then the radar system sends 2048 bits. Next comes the PN code marking the start of a new frame. On the ground you'd filter through the data till you recognize the PN code, then start counting bits so you can decommutate or separate out the data. The first 1024 bits are imaging, the next 512 bits are propulsion, and so on. However, this arrangement means that, for instance, radar always gets the same number of bits, even if the instrument has no data to transmit, which can be wasteful.

Rather than TDM, newer spacecraft use packetizing on both uplink and downlink in accordance with The Consultative Committee for Space Data Systems (CCSDS).[7] In the packetizing scheme for downlink, a burst of data called a packet comes from one instrument or subsystem, followed by a packet from another, and so on, in no specific order. The packets (and partial packets) are packed into the frame tightly, regardless of where each packet starts. Each packet carries an identification of the measurements it contains, the packet's size, etc., so the ground data system can recognize it and handle it properly. Data within a packet normally contains many different commutated measurements from the subsystem or instrument sending the packet. The spacecraft gathers packets (or perhaps parts of packets, depending on their sizes) into frames for downlink, marked with PN code as described above.

Packetizing schemes adhere to the International Standards Organization (ISO)'s Open Systems Interconnection (OSI) protocol suite, which recommends how computers of various makes and models can communicate. The ISO OSI is distance independent, and holds for spacecraft light-hours away as well as between workstations in the same office.

Demultiplexing is done at the frame level for TDM, and at the packet level for packet data. In both TDM and packets, the data carries a numeric key to determine which of several predetermined sets of measurements is present. On the ground, the key determines which decom map (decommutation map) will be used to separate out each measurement into its "channel" described below.

Telemetry Lock

Once a DSN receiver has locked onto a downlink, symbols are decoded into bits, and the bits go to a telemetry system. Once the telemetry system recognizes frames reliably, it is said to be in "lock" on the data. The data flow is discussed further in Chapter 18; the intent here is to highlight the difference between receiver lock and telemetry lock.

Channelization

Once all the measurements are identified in the ground system, they are typically displayed as channels. For example,

- Channel T-1234 always shows the temperature of the hydrazine tank.

- Channel A-1234 always shows the speed of reaction wheel number three.

- Channel E-1234 always shows the main bus voltage.

- Channel C-1234 always shows the state of the command decoder.

- Channel P-1234 always shows the pressure in the oxidizer tank.

And so on.

These are hypothetical channel identifications, of course. Each spacecraft identifies them differently and may have thousands of different measurements. But the value of each specific measurement will always appear in the same place, in its associated channel. This scheme makes it possible to arrange sets of related data within separate windows on your computer screen. Some data is typically not assigned channels at all, but delivered "raw" instead. Such raw data may include imaging data streams or data from other science instruments. In addition to telemetry measurements from the spacecraft, monitor data is available from the Deep Space Network to show the performance of its various subsystems, for example antenna elevation or receiver lock status. Monitor channels are typically identified with the letter M, such as Channel M-1234.

Notes

[1] http://www.well.com/user/dmsml/wilson.html
[2] http://zone.ni.com/devzone/cda/tut/p/id/1320
[3] http://nmp.jpl.nasa.gov/ds1
[4] http://spaceplace.jpl.nasa.gov/en/kids/vgr_fact2.shtml
[5] http://home.netcom.com/~chip.f/viterbi/tutorial.html
[6] http://www331.jpl.nasa.gov/public/JPLtcodes.html
[7] http://www.ccsds.org

Chapter 11

Typical Onboard Systems

Objectives: Upon completion of this chapter you will be able to describe the role of typical spacecraft subsystems: structural, thermal, mechanical devices, data handling, attitude and articulation control, telecommunications, electrical power and distribution, and propulsion. You will be able to list advanced technologies being considered for use on future spacecraft.

Systems, Subsystems, and Assemblies

One might expect a system to comprise subsystems, and subsystems to contain assemblies as in the ideal hierarchy illustrated at right. For example, a spacecraft, also called a flight system (to differentiate it from the ground system), might contain a dozen subsystems including an attitude control subsystem, which itself might contain dozens of assemblies including for example three or four reaction wheel assemblies, celestial reference assemblies, and inertial reference assemblies, etc.

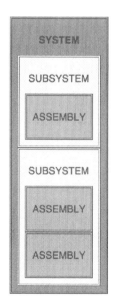

But all too often *system* and *subsystem* are used arbitrarily. In some usage a single system may comprise subsystems both aboard a spacecraft and on the ground, for example a telecommunications system with transmitter and receiver subsystems on both spacecraft and Earth. In other usage, as if to ensure permanent confusion of terms, frequently an instrument is named a subsystem, but it may contain lens systems, and so on.

Individual spacecraft can be very different from one another, and they can display different approaches to solving similar problems. Newer space-craft are smaller and less massive than their predecessors, yet there are common functions carried out by spacecraft regardless how massive or miniature.

Not all classifications of spacecraft have the same subsystems, though. An atmospheric probe spacecraft, for example, may lack propulsion and attitude control subsystems entirely. The discussions in this chapter mainly address subsystems that satisfy the requirements typical of complex flyby- or orbiter-class spacecraft, and in this way cover most simpler classes of spacecraft as well.

Subsystems discussed in this chapter include:

- Structure Subsystem
- Data Handling Subsystem
- Attitude & Articulation Control Subsystem
- Telecommunications Subsystem
- Electrical Power Subsystem
- Temperature Control Subsystem
- Propulsion Subsystem
- Mechanical Devices Subsystem
- Other Subsystems

Figures 11.1 and 11.2 may be useful as you proceed through this chapter and the next. They illustrate many components of the typical on-board systems, subsystems, and assemblies discussed in the text. The on-line version[1] offers links to further detail. In addition, there is a Galileo reference diagram available online.[2] Too large to be included in this book, it calls out similar components on the Galileo Jupiter orbiter and atmospheric probe spacecraft.

Structure Subsystem

The Structure subsystem provides overall mechanical integrity of the spacecraft. It must ensure that all spacecraft components are supported, and that they can withstand handling and launch loads as well as flight in freefall and during operation of propulsive components.

The spacecraft bus is a major part of a spacecraft's structure subsystem. It provides a place to attach components internally and externally, and to house delicate

Figure 11.3: Bus for Stardust S/C.

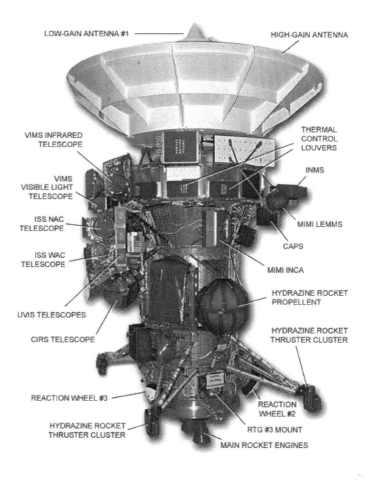

LOW-GAIN ANTENNA #1

HIGH-GAIN ANTENNA

VIMS INFRARED
TELESCOPE

THERMAL
CONTROL
LOUVERS

VIMS
VISIBLE LIGHT
TELESCOPE

INMS

ISS NAC
TELESCOPE

MIMI LEMMS

ISS WAC
TELESCOPE

CAPS

MIMI INCA

UVIS TELESCOPES

HYDRAZINE ROCKET
PROPELLENT

CIRS TELESCOPE

HYDRAZINE ROCKET
THRUSTER CLUSTER

REACTION WHEEL #3

REACTION
WHEEL #2

HYDRAZINE ROCKET
THRUSTER CLUSTER

RTG #3 MOUNT

MAIN ROCKET ENGINES

Figure 11.1: Cassini Spacecraft minus-Y side (without Huygens Probe).

modules requiring the protection of an environment with a measure of thermal and mechanical stability. It can provide an integral card chassis for supporting circuit boards of radio equipment, data recorders, and computers. It supports gyroscopes, reaction wheels, cables, plumbing, and many other components. The bus also influences the basic geometry of the spacecraft, and it provides the attachment points for other parts of the structure subsystem such as booms, antennas, and scan platforms. It also provides points that allow holding and moving the spacecraft during construction, testing, transportation, and launch.

A magnetometer boom appendage is typically the longest component of the structure subsystem on a spacecraft, although since it is deployable, it

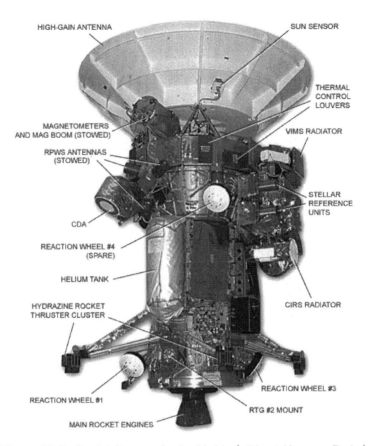

Figure 11.2: Cassini Spacecraft plus-Y side (without Huygens Probe).

may fall under the aegis of the mechanical devices subsystem discussed below. Since magnetometers (discussed in Chapter 12) are sensitive to electric currents near the spacecraft bus, they are placed at the greatest practical distance from them on a boom. The Voyager magnetometers are mounted 6 and 13 meters out the boom from the spacecraft bus. At launch, the mag boom, constructed of thin, non-metallic rods, is typically collapsed very compactly into a protective canister. Once deployed in flight, it cannot be retracted.

Data Handling Subsystems

Some of today's science instruments, or other subsystems, may easily have more embedded computing power than an entire Voyager spacecraft has. But there is usually one computer identified as the "spacecraft central" computer responsible for overall management of a spacecraft's activity. It is typically the same one which maintains timing, interprets commands from Earth, collects, processes, and formats the telemetry data to be returned to Earth, and manages high-level fault protection and safing routines. On some spacecraft, this computer is referred to as the command and data subsystem (CDS). For convenience, that term will be used here, recognizing that other names may apply to similar subsystems or sets of subsystems which accomplish some or all of the same tasks.

Sequence Storage

A portion of the CDS memory is managed as storage space for command sequences and programs uplinked from Earth that control the spacecraft's activities over a period of time. After use, these sequences and programs can be repeatedly overwritten with new ones to maintain optimum control of the spacecraft. These sequence loads are typically created by the project's planning and sequencing teams by negotiating and incorporating inputs from the spacecraft team, the science teams, and others.

Spacecraft Clock

As mentioned in Chapter 2, the spacecraft clock (SCLK, pronounced "sklock") is typically a counter maintained by CDS. It meters the passing of time during the life of the spacecraft. Nearly all activity within the spacecraft systems is regulated by the SCLK (an exception is realtime commands). The spacecraft clock may be very simple, incrementing every second and bumping its value up by one, or it may be more complex, with several main and subordinate fields that can track and control activity at multiple granularities. The SCLK on the Ulysses spacecraft, for instance, was designed to increment its single field by one count every two seconds. The Galileo and Magellan clocks, on the other hand, were designed as four fields of increasing resolution. Many types of commands uplinked to the spacecraft are set to begin execution at specific SCLK counts. In telemetry, SCLK counts that indicate data creation time mark engineering and science data whether it goes to the onboard storage device, or directly onto the downlink.

The presence of SCLK in telemetry facilitates processing, storage, retrieval, distribution, and analysis.

EVENT DESCRIPTION	SCET	S/C CLOCK
EXECUTE BOTH CAMERAS COMMAND ID: 450	335 13:28:45	1354282800:204
EXECUTE UV IMAGING SPECTROMETER OBSERVATION	335 13:28:45	1354282800:207

Figure 11.4: Excerpt from Cassini Sequence of Events during Jupiter phase December 2000. SCLK values for two commands appears next to their equivalent SCET times.

Telemetry Packaging and Coding

Telemetry data from science instruments and engineering subsystems is picked up by the CDS, where it is assembled into packages appropriate to the telemetry frame or packet scheme in use (for example CCSDS). If the spacecraft is down-

Figure 11.5: Cassini CDS main computer.

linking data in real time, the packet or frame may be sent to the spacecraft's transmitter. Otherwise, telemetry may be written to a mass storage device until transmission is feasible.

Spacecraft engineering or health data is composed of a wide range of measurements, from switch positions and subsystem states to voltages, temperatures, and pressures. Thousands of measurements are collected and inserted into the telemetry stream.

CDS's capability to alter the telemetry format and content accommodates various mission phases or downlink rates, as well as to enable diagnosis of anomalies. In the case of an anomaly, it may be necessary to temporarily terminate the collection of science data and to return only an enriched or specialized stream of engineering and housekeeping data.

Some data processing may take place within the CDS before science and engineering data are stored or transmitted. CDS may apply data compression methods to reduce the number of bits to be transmitted, and apply one or more encoding schemes to reduce data loss as described in Chapter 10.

Data Storage

It is rare for a mission to be provided the constant support of real-time tracking. Also, a spacecraft may spend time with its antenna off Earth-point while gathering data. For these reasons, spacecraft data handling subsystems are usually provided with one or more data storage devices such as tape recorders, or the solid-state equivalent of tape recorders which maintain large quantities of data in banks of memory, for example RAM or FLASH. The storage devices can be commanded to play out their stored data for downlink when DSN resources are available, and then to overwrite the old data with new.

Fault Protection and Safing

A robotic space flight system must have the intelligence and autonomy to monitor and control itself to some degree throughout its useful life at a great distance from Earth. Though ground teams also monitor and control the spacecraft, light time physically prohibits the ability to respond immediately to anomalies on the spacecraft. Tightly constrained tracking schedules also limit the ability to detect problems and respond. Fault protection (FP) algorithms, which normally run in one or more of the spacecraft's subsystems, therefore must ensure the ability to mitigate the impact of a mishap, and to re-establish the spacecraft's ability to contact Earth if an anomaly has caused an interruption in communications. A spacecraft may have many different FP algorithms running simultaneously with the ability to request CDS to take action.

Safing is one response that FP routines can request. Safing involves shutting down or reconfiguring components to prevent damage either from within or from the external environment. Another internal response may be an automated, methodical search to re-establish Earth-pointing and re-gain communications. This routine may or may not be a normal part of safing. While entrance into safing may temporarily disrupt ongoing science observations and require the flight team to perform additional work, safing provides strong and reliable protection for the spacecraft and its mission.

Usually a minimal set of safing-like instructions is also installed in ROM (it was contained in 1 kbyte on Magellan) where it can hide from even the worst imaginable scenarios of runaway program execution or power outage. More intricate safing routines (also called "contingency modes") and FP routines typically reside in CDS RAM, as well as parameters for use by the ROM code, where they can be updated as necessary during the life of the mission.

One example of a fault-protection routine is the Command-Loss Timer, CLT. This is a software timer running in CDS that is reset to a predetermined value, for example a week, every time the spacecraft receives a command from Earth. If the timer decrements all the way to zero, the assumption is that the spacecraft has experienced a failure in its receiver or other components in the command string. The CLT fault protection response issues commands for actions such as swapping to redundant hardware in an attempt to re-establish the ability to receive commands.

Attitude and Articulation Control Subsystems (AACS)

The path a rocket or guided missile takes during powered flight is directly influenced by its attitude, that is its orientation in space. During the atmospheric portion of flight, fins may deflect to steer a missile. Outside the atmosphere, changing the direction of thrust by articulating exhaust nozzles or changing the rocket's attitude influences its flight path. Thus the term guidance and control has become associated with attitude control during the powered ascent phase of a spacecraft's mission. After a few minutes' launch, though, a spacecraft may face a mission of many years in freefall, during which its attitude has no relation to guidance except during short, infrequent propulsive maneuvers.

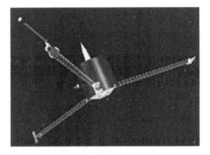

Figure 11.6: Spin-stabilized Lunar Prospector.

A spacecraft's attitude must be stabilized and controlled so that its high-gain antenna may be accurately pointed to Earth for communications, so that onboard experiments may accomplish precise pointing for accurate collection and subsequent interpretation of data, so that the heating and cooling effects of sunlight and shadow may be used intelligently for thermal control, and also for guidance: short propulsive maneuvers must be executed in the right direction.

Spin Stabilization

Stabilization can be accomplished by setting the vehicle spinning, like the Pioneer 10 and 11 spacecraft in the outer solar system, Lunar Prospector, and the Galileo Jupiter orbiter spacecraft, and its atmospheric probe (see page 137). The gyroscopic action of the rotating spacecraft mass is the stabilizing mechanism. Propulsion system thrusters are fired only occasionally to make desired changes in spin rate, or in the spin-stabilized attitude. In the case of Galileo's Jupiter atmospheric probe, and the Huygens Titan probe, the proper attitude and spin are initially imparted by the mother ship.

Active three-axis stabilization

Alternatively, a spacecraft may be designed for active 3-axis stabilization. One method is to use small propulsion-system thrusters to incessantly nudge the spacecraft back and forth within a deadband of allowed attitude error. Voyagers 1 and 2 have been doing that since 1977, and have used up a little over half their 100 kg of propellant as of April 2006. Thrusters are also referred to as mass-expulsion control systems, MEC, or reaction-control systems, RCS.

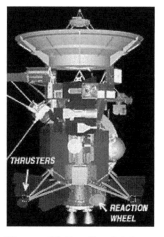

Another method for achieving 3-axis stabilization is to use electrically-powered reaction wheels, also called momentum wheels. Massive wheels are mounted in three orthogonal axes aboard the spacecraft.[3] They provide a means to trade angular momentum

Figure 11.7: Cassini is actively stabilized on 3 axes.

back and forth between spacecraft and wheels. To rotate the vehicle in one direction, you spin up the proper wheel in the opposite direction. To rotate the vehicle back, you slow down the wheel. Excess momentum that builds up in the system due to external torques, caused for example by solar photon pressure or gravity gradient, must be occasionally removed from the system by applying torque to the spacecraft, and allowing the wheels to acquire a desired speed under computer control. This is done during maneuvers called momentum desaturation, (desat), or momentum unload maneuvers. Many spacecraft use a system of thrusters to apply the torque for desats. The Hubble Space Telescope, though, has sensitive optics that could be contaminated by thruster exhaust, so it used magnetic torquers that interact with the Earth's magnetic field during its desat maneuvers.

There are advantages and disadvantages to both spin stabilization and 3-axis stabilization. Spin-stabilized craft provide a continuous sweeping motion that is desirable for fields and particles instruments, as well as some optical scanning instruments, but they may require complicated systems to de-spin antennas or optical instruments that must be pointed at targets for science observations or communications with Earth.

Three-axis controlled craft can point optical instruments and antennas without having to de-spin them, but they may have to carry out special rotating maneuvers to best utilize their fields and particle instruments. If thrusters are used for routine stabilization, optical observations such as imaging must be designed knowing that the spacecraft is always slowly rocking back and forth, and not always exactly predictably. Reaction wheels provide a much steadier spacecraft from which to make observations, but they add mass to the spacecraft, they have a limited mechanical lifetime, and they require frequent momentum desaturation maneuvers, which can perturb navigation solutions because of accelerations imparted by their use of thrusters.

No matter what choices have been made spin or 3-axis stabilization, thrusters or reaction wheels, or any combinations of these the task of attitude and articulation control falls to an AACS computer running highly evolved, sophisticated software.

Articulation

Many spacecraft have components that require articulation. Voyager and Galileo, for example, were designed with scan platforms for pointing optical instruments at their targets largely independently of spacecraft orientation. Many spacecraft, such as Mars orbiters, have solar panels which must track the Sun so they can provide electrical power to the spacecraft. Cassini's main engine nozzles are steerable. Knowing where to point a solar panel, or scan platform, or a nozzle that is, how to articulate it requires knowledge of the spacecraft's attitude. Since AACS keeps track of the spacecraft's attitude, the Sun's location, and Earth's location, it can compute the proper direction to point the appendages. It logically falls to one subsystem, then, to manage both attitude and articulation. The name AACS may even be carried over to a spacecraft even if it has no appendages to articulate.

Celestial Reference

Which way is "up"? Many different devices may be chosen to provide attitude reference by observing celestial bodies. There are star trackers, star scanners, solar trackers, sun sensors, and planetary limb sensors and trackers.[4]

Many of today's celestial reference devices have a great deal of intelligence, for example automated recognition of observed objects based on built-in star catalogs. Voyager's AACS takes input from a sun sensor for yaw and pitch reference. Its roll reference comes from a star tracker trained continuously on a single bright star (Canopus) at right angles to sunpoint. Galileo took its references from a star scanner which rotated with the spinning part of the spacecraft, and a sun sensor was available for use in ma-

Figure 11.8: Cassini Stellar Reference Unit, SRU.

neuvers. Magellan used a star scanner to take a fix on two bright stars during a special maneuver once every orbit or two, and its solar panels each held a sun sensor.

Inertial Reference

Celestial references are not always available or appropriate. Gyroscopes of some kind are typically carried as inertial reference devices to provide attitude reference signals to AACS for those periods when celestial references are not being used. For Magellan, this was the case nearly continuously; celestial references were used only during specific star scan maneuvers once every orbit or two. Other spacecraft, such as Galileo, Voyager and Cassini, were designed to use celestial reference nearly continuously, and they rely on inertial reference devices for their attitude reference only during relatively short maneuvers when celestial reference is lost.

In either case, gyro data must be taken with a grain of salt; mechanical gyroscopes precess and drift due to internal friction. Non-mechanical "gyros" also drift due to design constraints. Their rates of drift are carefully calibrated, so that AACS may compensate as best it can, when it computes its attitude knowledge using gyro references. (Note that the word "gyro" is commonly used as shorthand for "inertial reference unit," even though some units do not employ the gyroscopic effect at all.)

A few different inertial reference technologies are in use on today's spacecraft. The newer technologies depart from employing the mechanical gyroscopic effect, in favor of different physical principles:

Mechanical gyros Used on Voyager and Magellan, these rely on the rigidity in space of the axis of a spinning mass to provide attitude reference signals. This effect is easily demonstrated in a toy gyroscope. Mechanical gyros have limited lifespans due to mechanical wear.

Laser ring gyros and fiber-optic laser gyros: These use interferometry to sense the Doppler effect induced in beams of light, when the unit is rotated. Laser gyros have no moving parts to wear out.

Hemispherical resonator gyros: Used on NEAR and Cassini, these sense movement of a standing mechanical wave in a fused-silica shell. The wave is not unlike a wineglass 'singing' as you slide your finger around the rim. Null points in the wave precess when the unit is rotated. Other than their vibrating sensor shells, hemispherical resonator gyros have no moving parts.[5]

Don't Confuse Gyros and Reaction Wheels.
Gyros provide inertial reference inputs to AACS computers.
If they have any moving parts, they are small and lightweight.
Reaction wheels are fairly massive attitude control devices
at the output of AACS computers.

Telecommunications Subsystems

This section deals specifically with telecommunications equipment on board a spacecraft. A broader view of telecommunications issues may be found in Chapter 10, and the Deep Space Network's role is detailed in Chapter 18.

Telecommunications subsystem components are chosen for a particular spacecraft in response to the requirements of the mission profile. Anticipated maximum distances, planned frequency bands, desired data rates, available on-board electrical power (especially for a transmitter), and mass limitations, are all taken into account. Each of the components of this subsystem is discussed below:

High-Gain Antennas (HGA)

Dish-shaped HGAs are the spacecraft antennas principally used for long-range communications with Earth. The amount of gain achieved by an antenna (indicated here as high, low, or medium) refers to the relative amount of incoming radio power it can collect and focus into the spacecraft's receiving subsystems from Earth, and outbound from the spacecraft's transmitter.

In the frequency ranges used by spacecraft, this means that HGAs incorporate large paraboloidal reflectors. The cassegrain arrangement, described in Chapter 6, is the HGA configuration used most frequently aboard interplanetary spacecraft. Ulysses, which uses a prime focus feed, is one exception.

Figure 11.9: Ulysses prime-focus HGA.

HGAs may be either steerable or fixed to the spacecraft bus. The Magellan HGA, which also served as a radar antenna for mapping and as a drogue for aerobraking, was not articulated; the whole spacecraft had to be maneuvered to point the HGA to Earth for communications.

The Mars Global Surveyor spacecraft's HGA is on an articulated arm to allow the antenna to remain on Earth-point, independent of the spacecraft's attitude while it maps the surface of Mars. Galileo's HGA was designed to unfold like an umbrella after launch to enable the use of a larger diameter antenna than would have fit in the Space Shuttle cargo bay if a fixed antenna had been chosen. It did not unfurl, though, and Galileo had to carry out its mission using a low-gain antenna constrained to low data rates.

Figure 11.10: MGS's articulated HGA.

The larger the collecting area of an HGA, the higher the possible gain, and the higher the data rate it can support. The higher the gain, the more highly directional it is (within limits of frequency band in use). An HGA on an interplanetary spacecraft must be pointed to within a fraction of a degree of Earth for communications to be feasible. Once this is achieved, communications take place over the highly focused radio signal. This is analogous to using a telescope, which provides magnification (gain) of a weak light source, but requires accurate pointing.

To continue the telescope analogy, no magnification is achieved with the unaided eye, but it covers a very wide field of view, and need not be pointed with great accuracy to detect a source of light, as long as it is bright enough. In case AACS fails to be able to point a spacecraft's HGA with high accuracy for one reason or another, there must be some other means of communicating with the spacecraft.

An HGA can serve additional purposes unrelated to telecommunications. It can be a sunshade. Magellan's and Cassini's are good examples. They produced shade for the rest of the spacecraft when pointed to the

Sun. Cassini's HGA also served to protect the rest of the spacecraft from the thousands of micrometeoroid impacts it endured crossing Saturn's ring plane during Saturn Orbit Insertion.[6]

Low-Gain Antennas (LGA)

Low-gain antennas (LGAs) provide wide-angle coverage (the "unaided-eye," in the analogy) at the expense of gain. Coverage is nearly omnidirectional, except for areas which may be shadowed by the spacecraft body. LGAs are designed to be useable for relatively low data rates, as long as the

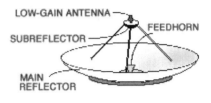

Figure 11.11: LGA atop HGA.

spacecraft is within relatively close range, several AU for example, and the DSN transmitter is powerful enough. Magellan could use its LGA at Venus's distance, but Voyager must depend on its HGA since it is over 90 AU away.

Some LGAs are conveniently mounted atop the HGA's subreflector, as in the diagram above. This was the case with Voyager, Magellan, Cassini, and Galileo. Ulysses carries an LGA atop its prime focus feed. A second LGA, designated LGA-2, was added to the Galileo spacecraft in the mission redesign which included an inner-solar system gravity assist. LGA-2 faced aft, providing Galileo with nearly full omnidirectional coverage by accommodating LGA-1's blind spots. Cassini's LGA-2 also is mounted near the aft end of the spacecraft, so it can provide coverage when LGA-1 is shaded by the spacecraft body.

Medium-Gain Antennas

MGAs are a compromise, providing more gain than an LGA, with wider angles of pointing coverage than an HGA, on the order of 20 or 30 degrees. Magellan carried an MGA consisting of a large cone-shaped feed horn, which was used during some maneuvers when the HGA was off Earth-point.

Spacecraft Transmitters

A spacecraft transmitter is a lightweight electronic device that generates a tone at a single designated radio frequency (see page 93), typically in the S-band, X-band, or Ka-band for communications and radio science. This tone is called the carrier. As described in Chapter 10, the carrier can be sent from the spacecraft to Earth as a pure tone, or it can be modulated with data or a data-carrying subcarrier.

The signal generated by the spacecraft transmitter is passed to a power amplifier, where its power is boosted to the neighborhood of tens of watts. This microwave-band power amplifier may be a solid state amplifier (SSA) or a traveling wave tube (TWT, also TWTA, pronounced "tweeta," for TWT Amplifier).

A TWTA is a vacuum tube. It uses the interaction between the field of a wave propagated along a waveguide, and a beam of electrons traveling along with the wave. The electrons tend to travel slightly faster than the wave, and on the average are slowed slightly by the wave. The effect amplifies the wave's total energy. TWTAs require a regulated source of high voltage. The output of the power amplifier is ducted through waveguides and commandable waveguide switches to the antenna of choice: HGA, MGA, or LGA.

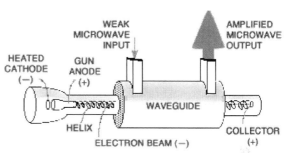

Figure 11.12: Travelling-Wave Tube Amplifier, TWTA.

Spacecraft Receivers

Commandable waveguide switches are also used to connect the antenna of choice to a receiver. The receiver is an electronic device which is sensitive to a narrow band of radio frequencies, generally plus and minus a few kHz of a single frequency selected during mission design. Once an uplink is detected within its bandwidth, the receiver's phase-lock-loop circuitry (PLL) will follow any changes in the uplink's frequency within its bandwidth. JPL invented PLL technology in the early 1960s, and it has since become standard throughout the telecommunications industry.

The receiver's circuitry can provide the transmitter with a frequency reference coherent with the received uplink (see page 147). This means the downlink signal's phase bears a constant relation to that of the uplink signal. The received uplink, once detected, locked onto, and stepped down in frequency, is stripped of its command-data-carrying subcarrier, which is passed to circuitry called a command detector unit (CDU). The CDU converts the analog phase-shifts that were modulated onto the uplink, into binary 1s and

0s, which are then typically passed to the spacecraft's Command and Data Subsystem (CDS) or equivalent.

Usually, transmitters and receivers are combined into a single electronic device on a spacecraft, which is called a transponder. Note that in everyday terrestrial use, a transmitter-receiver not designed to generate a coherent signal is called a transceiver.

Communications Relay

Mars orbiting spacecraft are equipped with a UHF antenna and receiver, to serve as a relay from spacecraft on the surface. The Mars Exploration Rovers Spirit and Opportunity, for example, routinely transmit their data in the UHF radio frequency band, as Mars Global Surveyor (MGS) passes overhead in Mars orbit. MGS stores the data on board, and then transmits it back to Earth on its X-band communications link. The same service is also provided by orbiters Mars Odyssey and Mars Express. There's more about relay operations in Chapter 17.

Electrical Power Supply and Distribution Subsystems

A quick primer on voltage, current, power, AC and DC appears at the end of this chapter; see page 184.

On today's interplanetary spacecraft, roughly between 300 W and 2.5 kW of electrical power is required to supply all the computers, radio transmitters and receivers, motors, valves, data storage devices, instruments, hosts of sensors, and other devices. Cassini uses roughly 1 kW. A power supply for an interplanetary spacecraft must provide a large percentage of its rated power over a lifetime measured in years or decades.

How does a spacecraft meet its demanding electrical power needs? Choices of technology to meet these requirements are today constrained largely to two: photovoltaics (PV) and radioisotope power systems (RPS). The latter includes radioisotope thermoelectric generators (RTGs) such as those currently powering Voyager and Cassini. Primary battery power is an option for use only on short-lived missions such as the Galileo or Huygens atmospheric-entry probes.

Photovoltaics

As the term suggests, photovoltaic materials have the ability to convert light directly to electricity. An energy conversion efficiency of about 30% was achieved in July 2000, and gains of up to ten percent more may be possible in coming years. Crystalline silicon and gallium arsenide are typical choices of materials for deep-space applications. Gallium arsenide crystals are grown especially for photovoltaic use, but silicon crystals are available in less-expensive standard ingots which are produced mainly for consumption in the microelectronics industry.

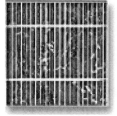

When exposed to direct sunlight at 1 AU, a current of about one ampere at 0.25 volt can be produced by a 6 cm diameter crystalline silicon solar cell. Gallium arsenide is notably tougher and more efficient. Amorphous silicon, less expensive and less efficient than crystalline, is employed in ultra-thin layers for residential and commercial PV applications.

Figure 11.13: Solar (photovoltaic) cell.

To manufacture spacecraft-grade solar cells, crystalline ingots are grown and then sliced into wafer-thin discs, and metallic conductors are deposited onto each surface: typically a thin grid on the sun-facing side and a flat sheet on the other. Spacecraft solar panels are constructed of these cells trimmed into appropriate shapes and cemented onto a substrate, sometimes with protective glass covers. Electrical connections are made in series-parallel to determine total output voltage. The resulting assemblies are called solar panels, PV panels, or solar arrays. The cement and the substrate must be thermally conductive, because in flight the cells absorb infrared energy and can reach high temperatures, though they are more efficient when kept to lower temperatures.

Figure 11.14: Mars Global Surveyor solar arrays.

Farther than about the orbit of Mars, the weaker sunlight available to power a spacecraft requires solar panel arrays larger than ever before. The Juno spacecraft,[7] due to launch in 2011 for Jupiter, is pushing this envelope. Magellan and Mars Observer were designed to use solar power, as was Deep Space 1, Mars Global Surveyor, Mars Pathfinder and, Sojourner, the Mars Exploration Rovers, Dawn, and Lunar Prospector. Topex/Poseidon, the Hubble Space Telescope, and most other Earth-orbiters also use solar power.

Solar panels typically have to be articulated to remain at optimum sun point, though they may be off-pointed slightly for periods when it may be desirable to generate less power. Spinning spacecraft may have solar cells on all sides that can face the Sun (see Lunar Prospector[8]). Prolonged exposure to sunlight causes photovoltaics' performance to degrade in the neighborhood of a percent or two per year, and more rapidly when exposed to particle radiation from solar flares.

In addition to generating electrical power, solar arrays have also been used to generate atmospheric drag for aerobraking operations. Magellan did this at Venus, as did MGS at Mars. Gold-colored aerobraking panels at the ends of MGS's solar arrays, visible in the image above, added to the aerodynamic drag for more efficient aerobraking.

Radioisotope Power Systems

RPSs enable, or significantly enhance, missions to destinations where inadequate sunlight, harsh environmental conditions, or operational requirements make other electrical power systems infeasible.

RPSs offer the key advantage of operating continuously, independent of unavoidable variations in sunlight. Such systems can provide power for long periods of time and at vast distances from the Sun. Additionally, they have little sensitivity to temperature, radiation or other space environmental effects. They are ideally suited for missions involving autonomous, long duration operations in the most extreme environments in space and on planetary surfaces.

RPSs, as currently designed for space missions, contain an isotopic mixture of the radioactive element plutonium in the form of an oxide, pressed into a ceramic pellet. The primary constituent of these fuel pellets is the plutonium isotope 238 (Pu^{238}). The pellets are arranged in a converter housing where they function as a heat source due to the natural decay of the plutonium. RTGs convert the heat into electricity through the use of simple thermocouples, which make use of the Seebeck effect and have no moving parts. The heat source in an ASRG, Advanced Stirling Radioisotope

Generator now under development, drives a moving piston and generator, offering much greater efficiency in producing electricity.

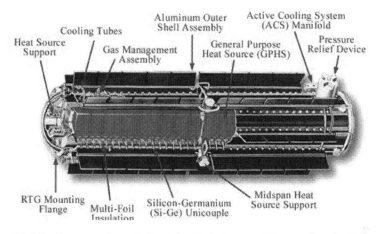

Figure 11.15: Cutaway view of Cassini's Radioisotope Thermoelectric Generators.

Since they remain thermally hot, RPSs present advantages and disadvantages. Cassini employs much of its RTGs' radiant heat inside its thermal blanketing, to warm the spacecraft and propellant tanks. On the other hand, RTGs must be located on the spacecraft in such a way to minimize their impact on infrared detecting science instruments. Galileo's RTGs are mounted behind shades to hide the near-infrared mapping spectrometer from their radiant heat. Shades are used on Cassini for similar reasons.

Figure 11.16: Red arrows show Cassini's RTG shades.

Seven generations of RPSs have been flown in space by the United States since 1961, powering 26 missions that have enabled world-renowned scientific exploration of the Moon, the Sun, Venus, Mars, Jupiter, Saturn, Uranus, Neptune, and—soon—Pluto. All of the RPSs on these historic solar system exploration missions have worked beyond their design lifetimes. An eighth RPS configuration, called the Multi-Mission Radioisotope Thermoelectric Generator (MMRTG), has recently been qualified for flight. It is planned for use on the Mars Science Laboratory rover, Curiosity.

Any NASA mission that proposes to use an RPS would undergo a comprehensive multi-agency environmental review, including public meetings and open comment periods during the mission planning and decision-making process, as part of NASA's compliance with the National Environmental Policy Act (NEPA). Additionally, any such mission proposed by NASA would not launch until formal approval for the missions's nuclear launch safety is received from the Office of the President.

Electrical Power Distribution

Virtually every electrical or electronic component on a spacecraft may be switched on or off via command. This is accomplished using solid-state or mechanical relays that connect or disconnect the component from the common distribution circuit, called a main bus. On some spacecraft, it is necessary to power off some set of components before switching others on in order to keep the electrical load within the limits of the supply. Voltages are measured and telemetered from the main bus and a few other points in the electrical system, and currents are measured and telemetered for many individual spacecraft components and instruments to show their consumption.

Typically, a shunt-type regulator maintains a constant voltage from the power source. The voltage applied as input to the shunt regulator is generally variable but higher than the spacecraft's required constant bus voltage. The shunt regulator converts excess electrical energy into heat, most of which is radiated away into space via a radiating plate. On spacecraft equipped with articulating solar panels, it is sometimes possible, and desirable for reasons of spacecraft thermal control, to off-point the panels from the Sun to reduce the regulator input voltage and thus reduce the amount of heat generated by the regulator.

Electrical Power Storage

Spacecraft which use photovoltaics usually are equipped with rechargeable batteries that receive a charge from the main bus when the solar panels are in the sunlight, and discharge into the bus to maintain its voltage whenever the solar panels are shadowed by the planet or off-pointed during spacecraft maneuvers. Nickel-cadmium (Ni-Cad) batteries are frequently used. After hundreds of charge-discharge cycles, this type of battery degrades in performance, but may be rejuvenated by carefully controlled deep discharge and recharge, an activity called reconditioning.

Thermal Control Subsystems

An interplanetary spacecraft is routinely subjected to extremes in temperature. Both the
Galileo and Cassini spacecraft were designed
for primary missions deep in the outer solar system, but their gravity-assist trajectories kept them in close to the Sun for a long
time. They had to be able to withstand solar
effects much stronger than here on Earth as
they flew by Venus during early cruise. And
for their missions at Jupiter and Saturn, their
design had to protect them from extreme cold.

Passive thermal control is obtained with
multi-layer insulation, MLI, which is often the
most visible part of a spacecraft. White or
gold-colored thermal blankets reflect infrared,
IR, helping to protect the spacecraft from excess solar heating. Gold is a
very efficient IR reflector, and is used to shade critical components.

The image above right shows the Cassini spacecraft fitted with its multi-
layer insulation thermal blanketing. Their gold color results from a reflective
silvery aluminum coating behind sheets of amber colored Kapton material.
MLI reflects sunlight to shade the spacecraft against overheating, and retains
internal spacecraft heat to prevent too much cooling.

Figure 11.17: Messenger's
folded solar panels with OSRs.

Optical solar reflectors (OSRs, also called
Thermal Control Mirrors), which are quartz
mirror tiles, are used for passive thermal control on some spacecraft, to reflect sunlight
and radiate IR. They were used extensively on
Magellan, operating at Venus's distance from
the Sun (0.72 AU). Messenger, the Mercury
Orbiter spacecraft, which has to endure operating only 0.38 AU from the Sun, employs
70% OSRs and 30% solar cells on its solar
panels, seen folded in the image at left. Its
sunshade, made of ceramic cloth, will keep the
spacecraft at room temperature despite sunlight eleven times more intense at Mercury's
distance than at Earth's.[9]

Active thermal control components include autonomous, thermostatically controlled resistive electric heaters, as well as electric heaters that can be commanded on or off from Earth. Electrical equipment also contributes heat when it is operating. When the equipment is turned off, sometimes a replacement heater is available to be turned on, keeping the equipment within its thermal limits. Sometimes radioisotope heater units (RHUs) are placed at specific locations on the spacecraft. Temperature sensors are placed at many locations throughout the spacecraft, and their measurements are telemetered to thermal engineers. They can command heaters as needed, and recommend any needed modifications to spacecraft operations to make sure no thermal limits are violated.

Louvers, which on some spacecraft may be counted under the mechanical devices subsystem, help minimize electrical power used for heaters to maintain temperature. They can help adapt to changes in the environment. Louvers vary the angle of their blades to provide thermal control by changing the effective emission of a covered surface. The louvers are positioned by bi-metallic strips similar to

Figure 11.18: Thermal control louvers.

those in a thermostat. They directly force the louvers open when internal temperatures are high, permitting heat to radiate into space. Cold internal temperatures cause the louvers to drive closed to reflect back and retain heat.

Active cooling systems, such as refrigeration, are generally not practical on interplanetary spacecraft. Instead, painting, shading, and other techniques provide efficient passive cooling. Internal components will radiate more efficiently if painted black, helping to transfer their heat to the outside.

For an atmospheric spacecraft, the searing heat of atmospheric entry is typically controlled by an aeroshell whose surface may be designed to ablate or simply to insulate with high efficiency. After entering the atmosphere, the aeroshell is typically jettisoned to permit the spacecraft to continue its mission.

Micro-meteoroid Protection

Multi-layer thermal insulation blankets also provide some protection against micro-meteoroid impacts. They are made with Kapton, Kevlar, or other fabrics strong enough to absorb energy from high-velocity micro-meteoroids before they can do any damage to spacecraft components. Impact hazards are greatest when crossing the ring planes of the Jovian planets.

Voyager recorded thousands of hits in these regions, as did Cassini crossing Saturn's ring plane, fortunately from particles no larger than smoke particles. Spacecraft sent to comets, such as Giotto and Stardust, carry massive shields to protect from hits by larger particles.

Figure 11.19: Stardust spacecraft with shields.

Propulsion Subsystems

Launch vehicles, addressed in Chapter 14, provide the massive propulsion needed to leave Earth and start out on an interplanetary trajectory. Once the few minutes of launch acceleration are over, a spacecraft uses its smaller-capacity on-board propulsive devices to manage 3-axis stability or control its spin, to execute maneuvers, and to make minor adjustments in trajectory. The more powerful devices are usually called engines, and they may provide a force of several hundred newtons. These may be used to provide the large torques necessary to maintain stability during a solid rocket motor burn, or they may be the only rockets used for orbit insertion.

Smaller devices, generating between less than 1 N and 10 N, are typically used to provide the delta-V (ΔV) for interplanetary trajectory correction maneuvers, orbit trim maneuvers, reaction wheel desaturation maneuvers, or routine 3-axis stabilization or spin control. Many of the activities of propulsion subsystems are routinely initiated by AACS. Some or all may be directly controlled by or through CDS.

Figure 11.20: Magellan rocket engine module.

Figure 11.20 is a photo showing one of the Magellan spacecraft's four rocket engine modules, which were mounted on struts not unlike Cassini's. Each module has two 445-N engines, one gold-colored 22-N thruster, and three gold colored 1-N thrusters. The 445-N engines were aimed aft for large mid-flight course corrections and orbit-trim corrections, and for controlling the spacecraft while its solid rocket motor burned during Venus orbit insertion. The 22-N thrusters kept the spacecraft from rolling during those same maneuvers. The 1-N thrusters were used for momentum wheel desaturation and other small maneuvers. Additional examples of thrusters can be found at this supplier's website.

Other components of propulsion subsystems include propellant tanks, plumbing systems with electrically or pyrotechnically operated valves, and helium tanks to supply pressurization for the propellant. Some propulsion subsystems, such as Galileo's, use hypergolic propellants—two compounds stored separately which ignite spontaneously upon being mixed in the engines or thrusters. Other spacecraft use hydrazine, which decomposes explosively when brought into contact with an electrically heated metallic catalyst within the engines or thrusters. Cassini, whose propulsion system is illustrated in Figure 11.21, uses both hypergolics for its main engines and hydrazine monopropellant for its thrusters.[10]

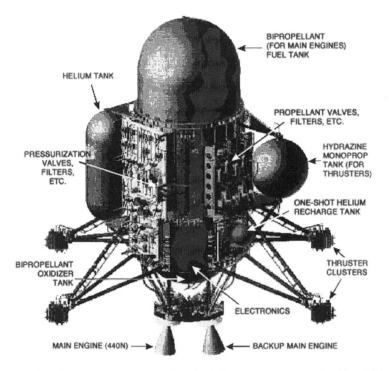

Figure 11.21: Cassini's Propulsion Module Subsystem, courtesy Lockheed Martin.

The Deep Space 1 spacecraft[11] was a pioneer in the use of ion-electric propulsion in interplanetary space. With their high specific impulse (due to high nozzle exit velocities), ion engines can permit spacecraft to achieve the high velocities required for interplanetary or interstellar flight.

An ion engine functions by taking a gas such as xenon, and ionizing it (removing electrons from the atoms) to make it responsive to electric and magnetic fields. Then the ions are accelerated to extremely high velocity using electric fields and ejected from the engine. Electrical power comes from arrays of photovoltaic cells converting sunlight, so the technology is also called solar-electric propulsion. The act of ejecting mass at extremely high speed provides the classical action for which the reaction is spacecraft acceleration in the opposite direction. The much higher exhaust speed of the ions compared to chemical rocket exhaust is the main factor in the engine's higher performance.[12]

The ion engine also emits electrons, not to take advantage of accelerating their tiny mass, but to avoid building a negative electric charge on the spacecraft and causing the positively charged ion clouds to follow it.

Mechanical Devices Subsystems

A mechanical devices subsystem typically supplies equipment to a spacecraft that deploys assemblies after launch. Often it provides motion that can be initiated, but once initiated is not controlled by feedback or other means. Some of these are pyrotechnically initiated mechanisms (pyros). These devices may be used to separate a spacecraft from its launch vehicle, to permanently deploy booms, to release instrument covers, to jettison an aeroshell and deploy parachutes, to control fluid flow in propulsion and pressurization systems, and to perform many other such functions.

While pyro devices are lightweight, simple and reliable, they have drawbacks including their "one-shot" nature, mechanical shock, and potential hazards to people handling them. Mars Pathfinder depended on 42 pyro device events during atmospheric entry, descent and landing. Each Mars Exploration Rover spacecraft, Spirit and Opportunity, fired 126 pyro devices during Mars atmosphere entry, descent, and landing. A launch vehicle may depend on a hundred or more pyro events during ascent.

Prior to initiating or "setting off" a pyro device, an assembly in the electrical power subsystem, typically called a pyrotechnic switching unit (PSU), may be commanded to operate, charging a capacitor bank that can provide the spike of high current the device needs to fire, protecting the main bus from a momentary power drain. Pyro devices typically used on a spacecraft include pyro valves, "explosive" bolts, zip cord, cable cutters, and pin pullers. "Explosive" bolt is a misnomer since modern devices are designed to separate with little mechanical shock and no stray particle release.

Alternative technologies are becoming available to perform some of the tasks traditionally performed by pyro devices. For example paraffin actuators perform mechanical functions when electrically heated.

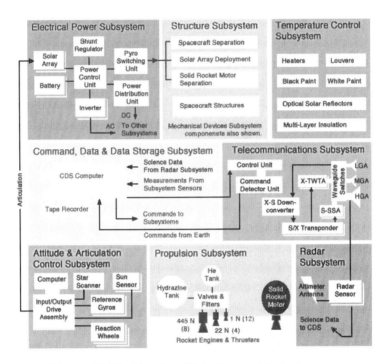

Figure 11.22: Magellan Flight System block diagram.

Figure 11.22 is a block diagram that illustrates how an interplanetary Space Flight System functions, by referring to the Magellan spacecraft, which combined many of the subsystems discussed in this chapter. Magellan's mission succeeded in obtaining high resolution data from the entire surface of Venus in the early 1990's. The spacecraft carried only one science instrument, the Radar, to penetrate Venus's cloud cover. Otherwise, the spacecraft's subsystems are representative of those found on many other spacecraft. Boxes within the diagram are shown double or triple to indicate the presence of two or three units of the same name. The numbers 12, 4, and 8 below the rocket engines indicate the quantity of each kind installed.

Redundancy and Flexibility

The hallmark of modern automated spacecraft is the ability to maintain or restore functionality after component failures. Components fail unexpectedly during the life of a mission. Most components upon which the success of the mission depends have redundant backups, and the means to reroute functional flow to accommodate their use either autonomously or via commanding in real time. Spacecraft such as Voyager, Galileo, Ulysses, and Pioneer, have enjoyed greatly extended missions, returning valuable science data long after their primary missions have been completed. This is due entirely to the on-board availability of redundant transmitters, receivers, tape recorders, gyroscopes, and antennas. The ability to modify on-board flight software has allowed them the flexibility to increase and extend functionality based on newly conceived techniques.

Advanced Technologies

Ongoing research at JPL and other institutions is producing new technologies useful for less costly and more capable, more reliable and efficient spacecraft for future space missions. Advances in such areas as spacecraft power, propulsion, communications, navigation, data handling systems, pointing control, and materials is expected to increase by many times the potential science returns from future missions. Here are a few of the many promising technologies being developed:

Solar sails, which use solar radiation pressure in much the same way that a sailboat uses wind, promise to provide the means for high-speed interplanetary or interstellar propulsion. JAXA's Ikaros spacecraft[13] successfully demonstrated solar sailing beginning in May, 2010.

Figure 11.23: Image of proposed solar sail by *Voiles Solaires.*

Telecommunications systems are being developed to operate in K and Ku bands, higher frequencies than the current S- and X-band systems.

Laser telecommunications systems are also being explored which modulate data onto beams of coherent light instead of radio.[14] Among the advantages to laser telecommunications are low power consumption, much higher data rates, and reduced-aperture Earth stations. The pointing requirements for laser communication are much more stringent than for microwave radio communication. During Galileo's Earth-2 flyby en route to Jupiter, JPL succeeded in transmitting laser signals to Galileo, which received them as points of light detected by the Solid State Imaging System (SSI).

Interferometry will be a technology that enables science observations to achieve immense improvements in spatial resolution and sensitivity. Space-

borne radio telescopes, infrared telescopes and visible-light telescopes will
be flown in exacting formation to synthesize large apertures for many different types of investigations in astronomy, astrophysics and cosmology. These
instruments are expected to eventually be able to image planets around
neighboring stars.

JPL has many planned and proposed missions which will take advantage
of advanced technologies.

Electrical Primer

Two basic electrical measurements are *voltage* and *current*.

We can compare electrical voltage with water pressure. The 12 volts (V)
in your automobile is a relatively low electrical "pressure." By analogy, low
water pressure could be used in, say, a drinking fountain. The 120-volt
service in United States homes provides a higher electrical "pressure." If
you were to increase the water pressure in your drinking fountain tenfold,
and the pipes didn't burst, the water might squirt across the room when
you went to take a sip.

Current is analogous to the water's flow rate. The small pipes in a drinking
fountain carry a small current of water — only a liter or so per minute.
Large pipes supplying a fire hydrant can carry a substantially higher "current," perhaps thousands of liters per minute.

You can vary the flow rate in your drinking fountain by changing the
pressure, and this is true in electrical circuits as well: increase the voltage supplied to a resistor such as a lightbulb, and more current will flow
through it. You could continue to increase the voltage until the increasing
current overloads and burns out the light bulb (bursts its "pipes").

Current, I, equals electromotive force (voltage) E, divided by resistance,
R. Current is measured in amperes (also called amps), A, electromotive
force in volts, V, and resistance in ohms, Ω.

$$I = E/R$$

If the voltage across a resistor increases, the current will also increase. If
the resistance increases, the current will decrease.

You wouldn't use drinking-fountain-sized pipes to fight a fire. Likewise,
it takes larger-diameter electrical wires to safely carry a larger electrical
current.

Electrical current is measured in amperes (A). The 12-volt fan in your automobile might consume 1 A at its low setting, and 4 A on high.

Volts multiplied by amperes gives a convenient single measure of power: watts, abbreviated W.

$$W = VA$$

Alternatively, watts are called volt-amps, VA.

If a current flows steadily in one direction, as it does right out of a battery, or RTG, or solar panel, it's called direct current, DC. If it flows back and forth as it does in most residential service, it's called alternating current, AC. Some components on a spacecraft may require AC, so they use an inverter to change DC to AC. If you need to change AC to DC, you'd use a rectifier. It's easy to step voltage either up or down if it's AC, using a simple wire-wound device called a transformer. DC can be stepped down in voltage, and that voltage held constant, using a voltage regulator. Stepping DC up to a higher-voltage DC requires an intermediate conversion to AC and rectification.

Notes

[1]http://www.jpl.nasa.gov/basics/cassini
[2]http://www.jpl.nasa.gov/basics/bsf11-5a.gif
[3]http://www.sti.nasa.gov/tto/spinoff1997/t3.html
[4]http://www.ballaerospace.com/page.jsp?page=104
[5]http://www.es.northropgrumman.com/solutions/hrg/
[6]http://www.jpl.nasa.gov/basics/soi
[7]http://www.nasa.gov/mission_pages/juno
[8]http://nssdc.gsfc.nasa.gov/planetary/lunarprosp.html
[9]http://messenger.jhuapl.edu/the_mission/spacecraft_design.html
[10]http://www.braeunig.us/space/propuls.htm#mono
[11]http://nmp.jpl.nasa.gov/ds1
[12]http://nmp.jpl.nasa.gov/ds1/tech/sep.html
[13]http://www.jspec.jaxa.jp/e/activity/ikaros.html
[14]http://lasers.jpl.nasa.gov

Chapter 12

Typical Science Instruments

Objectives: Upon completion of this chapter you will be able to distinguish between remote- and direct-sensing science instruments, state their characteristics, recognize examples of them, and identify how they are classified as active or passive sensors.

Most interplanetary missions are flown to collect science data. The exceptions are spacecraft like Deep Space 1, whose purpose is to demonstrate new technologies and validate them for future use on science missions. On a science mission, though, all the engineering subsystems and components that we've discussed up to this point serve a single purpose. That purpose is to deliver scientific instruments to their destination, to enable them to carry out their observations and experiments, and to return data from the instruments.

Science Payload

There are many different kinds of scientific instruments. They are designed, built, and tested by teams of scientists working at institutions around the world who deliver them to a spacecraft before its launch. Once delivered, they are integrated with the spacecraft, and tests continue with other subsystems to verify their commands function as expected, telemetry flows back from the instrument, and its power, thermal, and mechanical properties are within limits. After launch, the same scientists who created an instrument may operate it in flight through close cooperation with the rest of the flight team.

The purpose of this chapter is not to describe all science instruments. There are too many, they are complicated, and new ones are always being designed. We can, however, describe the basic categories of science instruments, and populate them with some examples. This will provide some keys so you will be able to recognize current and future science instruments, to immediately know their general purpose, to broadly understand how they operate, and to realize what their basic operational requirements must be.

Many science instruments are now described in detail on the Web, which makes it convenient to learn all about them. Some high quality links are provided in the chapter endnotes.

Direct- and Remote-sensing Instruments

Direct-sensing instruments interact with phenomena in their immediate vicinity, and register characteristics of them. The heavy ion counter that flew on Galileo is a direct sensing instrument. It registers the characteristics of ions in the spacecraft's vicinity that actually enter the instrument. It does not attempt to form any image of the ions' source. Galileo and Cassini each carry dust detectors. These measure properties, such as mass, species, speed, and direction, of dust particles which actually enter the instrument. They do not attempt to form any image of the source of the dust.

Figure 12.1: Galileo Heavy Ion Counter.

Remote-sensing instruments, on the other hand, exist to form some kind of image or characterization of the source of the phenomena that enter the instrument. In doing so, they record characteristics of objects at a distance, sometimes forming an image by gathering, focusing, and recording light. A camera, also called an imager, is a classic example of a remote-sensing instrument.

Active and Passive Instruments

Most instruments only receive and process existing light, particles, or other phenomena, and they are said to be passive. Typical of this type would be an imaging instrument viewing a planet that is illuminated by sunlight, or a magnetometer measuring existing magnetic fields.

An active instrument provides its own means of sensing the remote object. Typical of this would be a radar system. Radar generates pulses of radio waves that it sends to a surface, and then receives their reflections back from the surface, to create images or deduce characteristics of the surface. Some radio science experiments, described in Chapter 8, are also examples of active sensing, since they send radio energy through a planet's atmosphere or rings to actively probe the phenomena, with the results being received directly on Earth.

Most of the active-sensing instruments on a spacecraft are also remote-sensing instruments. An exception, an active direct-sensing instrument, would be one that comes in contact with an object of interest while providing a source of energy to probe the object. An example of this type of instrument is the Alpha-Particle X-ray Spectrometer, APXS, carried by the Sojourner Rover on Mars. The instrument contained a source (curium) of alpha-particle

Figure 12.2: APXS on Sojourner Rover.

radiation (helium nuclei) that the rover placed directly upon various sample rocks on the Martian surface, to determine their composition by acquiring energy spectra of the backscattered alpha particles and X-rays returned from the sample.

Examples of Direct-Sensing Science Instruments

High-energy Particle Detectors

High-energy particle detector instruments measure the energy spectra of trapped energetic electrons, and the energy and composition of atomic nuclei. They may employ several independent solid-state-detector telescopes. The Cosmic Ray Subsystem, CRS, on the Voyagers measures the presence and angular distribution of particles from planets' mag-

Figure 12.3: CRS on Voyager.

netospheres, and from sources outside our solar system: electrons of 3-110 MeV and nuclei 1-500 MeV from hydrogen to iron. The Energetic Particle Detector (EPD) on Galileo is sensitive to the same nuclei with energies from 20 keV to 10 MeV.

Low-Energy Charged-Particle Detectors

A low-energy charged-particle detector (LECP) is a mid-range instrument designed to characterize the composition, energies, and angular distributions of charged particles in interplanetary space and within planetary systems.

One or more solid-state particle detectors may be mounted on a rotating platform. The Voyagers' LECPs are sensitive from around 10 keV up into the lower ranges of the Cosmic Ray detector. Ulysses' LECP is similar, and is named GLG for its Principal Investigators Gloeckler and Geiss.

Plasma Instruments

Plasma detectors serve the low-end of particle ener-gies. They measure the density, composition, tem-perature, velocity and three-dimensional distribution of plasmas, which are soups of positive ions and electrons, that exist in interplanetary regions and within planetary magnetospheres. Plasma detectors are sensitive to solar and planetary plasmas, and they observe the solar wind and its interaction with

Figure 12.4: CAPS on Cassini.

a planetary system. The Cassini Plasma Spectrometer Subsystem, CAPS, measures the flux (flow rate or density) of ions as a function of mass per charge, and the flux of ions and electrons as a function of energy per charge and angle of arrival.

Dust Detectors

Dust detectors measure the number, velocity, mass, charge, and flight direction of dust particles strik-ing the instrument. As an example, Galileo's in-strument can register up to 100 particles per sec-ond and is sensitive to particle masses of between 10-16 and 10-6 gram. The Heidelberg Dust Re-search Group is responsible for dust detection ex-periments on many spacecraft, including Galileo, Ulysses, Cassini, DUNE, ISO, Rosetta, and Star-dust. Cassini's Cosmic Dust Analyzer, CDA, can determine the species of material in some dust parti-cles as well as the properties mentioned above. Some very informative animations of how it operates can be found here.

Figure 12.5: CDA on Cassini.

Magnetometers

Magnetometers are direct-sensing instruments that detect and measure the interplanetary and solar magnetic fields in the vicinity of the spacecraft.

They typically can detect the strength of magnetic fields in three planes. As a magnetometer sweeps an arc through a magnetic field when the spacecraft rotates, an electrical signature is produced proportional to the strength and structure of the field.

The Voyager Magnetometer Experiment (MAG) consists of two low-field magnetometers and two high-field magnetometers that together provide measurement of fields from 0.02 nano-Tesla (nT) to 2,000,000 nT. Voyager magnetometers were built at NASA GSFC where their investigation team resides. On the spacecraft, the instruments populate a 13-meter-long fiberglass boom to keep them away from on-board interference. The magnetometers provide

Figure 12.6: MAG Boom extends toward Voyager's upper left.

direct field measurement of both the planetary and the interplanetary media. Having completed their exploration of the outer planets, the Voyager magnetometers are now a key component of the Voyager Interstellar Mission, collecting measurements of magnetic fields far from the Sun.

It is typical for magnetometers to be isolated from the spacecraft on long extendable booms. This is a picture of Voyager's MAG boom being extended during a pre-launch test. Since the boom cannot support its own weight under the 1-G gravitational acceleration on Earth's surface, a cage arrangement suspends the boom as it extends. The 13-m boom was unfurled from a canister less than a meter in length.

The Cassini spacecraft deployed its magnetometer boom, supporting dual technique magnetometers, during the Earth-flyby period of its mission in 1999.

Plasma Wave Detectors

Plasma wave detectors typically measure the electrostatic and electromagnetic components of plasma waves in three dimensions. The instrument functions like a radio receiver sensitive to the wavelengths of plasma in the solar wind from about 10 Hz to about 60 kHz. When within a planet's magnetosphere, it can be used to detect atmospheric lightning and events when dust particles strike the spacecraft. The Voyagers' plasma wave data[1] has enabled sound recordings of the particle hits the spacecraft experienced passing through the ring planes of the outer planets.

Mass Spectrometers

Mass spectrometers, for example the Cassini Spacecraft's INMS, Ion and Neutral Mass Spectrometer,[2] report the species of atoms or molecules that enter the instrument. Another example is the Huygens' GCMS, Gas Chromatograph Mass Spectrometer,[3] which analyzed the atmosphere of Titan as the probe parachuted toward Titan's surface in 2004. After landing, it also reported on surface composition. A site called "Chemguide" provides some general information on how mass spectrometers work.[4]

Examples of Remote-Sensing Science Instruments

Planetary Radio Astronomy Instruments

A planetary radio astronomy instrument measures radio signals emitted by a target such as a Jovian planet. The instrument on Voyager is sensitive to signals between about 1 kHz and 40 MHz and uses a dipole antenna 10 m long, which it shares with the plasma wave instrument. The planetary radio astronomy instrument detected emissions from the heliopause in 1993 (see the illustration on page 24). Ulysses carries a similar instrument.

Imaging Instruments

An imaging instrument uses optics such as lenses or mirrors to project an image onto a detector, where it is converted to digital data. Natural color imaging requires taking three exposures

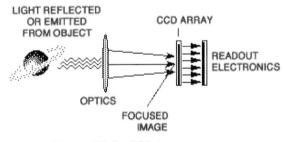

Figure 12.7: CCD Imaging system.

of the same target in quick succession through different color filters, typically selected from a filter wheel. Earth-based processing combines data from the three black and white images, reconstructing the original color by utilizing the three values for each picture element (pixel). Movies are produced by taking a series of images over an extended period of time.

This Cassini Imaging Science FAQ (Frequently Asked Questions list) for raw, unprocessed images, offers some insight into obtaining and processing images from spacecraft.[5]

In the past, the detector that created the image was a vacuum tube resembling a small CRT (cathode-ray tube), called a vidicon. In a vidicon, signals applied to deflection coils and focus coils sweep an electron beam from a heated cathode (electron source) across a photoconductor coating inside the tybe's glass front where the image is focused. Light striking the photoconductor causes its grains to leak their electric charge in proportion to the light's intensity. The sweeping beam's current would vary as it recharges depleted grains' charges to the point they repel the beam's electrons. This current becomes the basis for the digital video signal produced. Viking, Voyager, and Mariner spacecraft used vidicon-based imaging systems. A vidicon requires a fairly bright image to detect. The Voyagers' wide angle and narrow angle camera specifications make for some very informative reading.[6]

Modern spacecraft use CCDs, charge-coupled devices. A CCD is usually a large-scale integrated circuit having a two-dimensional array of hundreds of thousands, or millions, of charge-isolated wells, each representing a pixel. Light falling on a well is absorbed by a photoconductive substrate such as silicon and releases a quantity of electrons proportional to the intensity of light. The CCD detects and stores accumulated electrical charge representing the light level on each well over

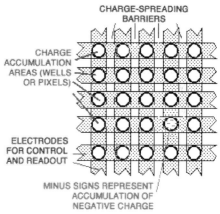

Figure 12.8: Detail of a CCD Detector.

time. These charges are read out for conversion to digital data. CCDs are much more sensitive to light of a wider spectrum than vidicon tubes, they are less massive, require less energy, and they interface more easily with digital circuitry. It is typical for CCDs to be able to detect single photons.

Galileo's Solid State Imaging instrument, SSI, which pioneered the technology, contains a CCD with an 800×800 pixel array. The optics for Galileo's SSI, inherited from Voyager, consist of a Cassegrain telescope with a 176.5-mm aperture and a fixed focal ratio of f/8.5. Since the SSI's wavelength range extends from the visible into the near-infrared, the experimenters are able to map variations in the satellites' color and reflectivity that show differences in the composition of surface materials.

Not all CCD imagers have two-dimensional arrays. The HIRISE instrument, or High-Resolution Imaging Science Experiment, on the Mars Reconnaissance Orbiter (MRO) spacecraft has detectors made of single lines of CCD sensors. A two-dimensional image is built up as the image of the Martian surface moves across this one-dimensional detector while the spacecraft moves in orbit, much as a page moves across the detector in a fax machine (this is called "pushbroom" mode). HIRISE often achieves a resolution of less than 25cm per pixel. The Mars Orbiter Camera (MOC) on the Mars Global Surveyor spacecraft, also used a single-dimension detector.

The Magnetosphere Imager

Cassini carries a unique instrument that's never been flown before in the outer solar system. The Magnetospheric Imaging Instrument (MIMI) Ion and Neutral Camera (INCA) can form images of the giant magnetic envelopes of Jupiter and its main objective Saturn, as well as fields associated with Saturn's moons. MIMI INCA doesn't use CCDs to make images. In fact it doesn't even use light at all. MIMI INCA is more like a particle detector, although unlike most particle detectors, it is actually a remote sensing instrument. MIMI INCA senses ions and neutral atoms that have been flung out of a planet's magnetosphere, forming an image of the source of the particles.

Polarimeters

The molecules of crystals of most materials are optically asymmetrical; that is, they have no plane or center of symmetry. Asymmetrical materials have the power to rotate the plane of polarization of plane-polarized light.

Polarimeters are optical instruments that measure the direction and extent of the polarization of light reflected from their targets. Polarimeters consist of a telescope fitted with a selection of polarized filters and optical detectors. Careful analysis of polarimeter data can infer information about the composition and mechanical structure of the objects reflecting the light, such as various chemicals and aerosols in atmospheres, rings, and satellite surfaces, since they reflect light with differing polarizations. A polarimiter's function may be integrated with another instrument, such as a camera, or the Voyager photopolarimeter that combines functions with a photometer.

Photometers

Photometers are optical instruments that measure the intensity of light from a source. They may be directed at targets such as planets or their satellites to quantify the intensity of the light they reflect, thus measuring the object's reflectivity or albedo. Also, photometers can observe a star while a planet's rings or atmosphere intervene during occultation, thus yielding data on the density and structure of the rings or atmosphere. One of the three instruments on the Spitzer Space Infrared Facility (SIRTF) is a photometer designed to measure the intensity of stars in the infrared.

Spectroscopic Instruments

Spectroscopy (see Chapter 6) provides a wealth of information for analysis of observed targets. The field of spectroscopy includes methods for analysis of virtually every part of the electromagnetic spectrum. Most are differentiated as either atomic or molecular based methods, and they can be classified according to the nature of their interaction:

- Absorption spectroscopy uses the range of the electromagnetic spectra in which a substance absorbs. Signatures of aroms or molecules can be recognized by dips in a spectrum's intensity at specific wavelengths, as light passes through an atmosphere for example.

- Emission spectroscopy uses the range of electromagnetic spectra in which a substance radiates. Hot metals, for example, radiate a continuous spectrum of many (or all) wavelengths, while excited gasses emit various discreet wavelengths, giving each specimen a characteristic "fingerprint."

- Scattering spectroscopy measures the amount of light that a substance scatters at certain wavelengths, incident angles, and polarization angles.

Spectroscopic instruments are optical remote-sensing instruments that split the light received from their targets into component wavelengths for measuring and analyzing. In general, a spectroscope provides spectral results visible to the eye, for example through an eyepiece on the instrument. Spectrometers and spectrographs use electronic detectors, such as CCDs, and return data representing the intensity of various parts of the spectrum observed. Instruments on spacecraft typically employ a diffraction grating to disperse the incoming light signal, while a spectroscope might employ either a prism or a grating.

A Compact Disc (CD) or Digital Versatile Disk (DVD) offers an example of a spectroscopic diffraction grating. Observing a bright light shining on the microscopic data tracks on its surface demonstrates the effect diffraction gratings produce, separating light into its various wavelength, or color, components. The apparent separation of wavelengths is due to constructive and destructive interference of light waves reflecting at varying angles from the grating lines according to their wavelengths. A similar effect can be seen from a thin layer of oil floating on water. Inexpensive "diffraction glasses" provide another effective way to explore spectroscopy at home or in the classroom.

Spectrometers carried on spacecraft are typically sensitive in the infrared, visible, and ultraviolet wavelengths. The Near-Infrared Mapping Spectrometer, NIMS, on Galileo maps the thermal, compositional, and structural nature of its targets using a two-dimensional array of pixels.

An imaging or mapping spectrometer provides spectral measurements for each of the many the pixels of an image the instrument obtains, thus supplying spectral data for many different points on a target body all at once.

Cassini's ultraviolet instrument is the Ultraviolet Imaging Spectrograph, UVIS. Its infrared instrument is the Composite Infrared Mapping Spectrometer, CIRS. Cassini's Visible and Infrared Mapping Spectrometer, VIMS, produce images whose every pixel contains spectral data at many different wavelengths. It is as though the instrument returns a whole stack of images with each observation, one image at each wavelength. Such data units are frequently called "cubes."

Infrared Radiometers

An infrared radiometer is a telescope-based instrument that measures the intensity of infrared (thermal) energy radiated by the targets. One of its many modes of observing is filling the field of view completely with the disc of a planet and measuring its total thermal output. This technique permits the planet's thermal energy balance to be computed, revealing the ratio of solar heating to the planet's internal heating. Another application measures Mars's atmospheric and surface thermal properties.

Combinations

As mentioned above, sometimes various optical instrumentation functions are combined into a single instrument, such as photometry and polarimetry combined into a photopolarimeter, or spectroscopy and radiom-

Figure 12.9: PPR on Galileo.

etry combined into a radiometer-spectrometer instrument. One example is Galileo's Photopolarimeter Radiometer instrument, PPR. Another example is the Voyagers' infrared interferometer spectrometer and radiometers, IRIS.

Cooling

CCD detectors in imagers and some spectrometry instruments perform best when they are cool. In infrared instruments, cooling essential for maintaining a signal-to-noise ratio that permits useful observations. Fortunately, it is easy to cool some detectors in flight. On Cassini, for example, each of the optical detectors is mounted to a thermally conductive metal part, that thermally connects to a radiator facing deep space. As long as flight operations keeps sunlight off the radiator, the detector's heat will radiate away into space. Detector temperatures in the neighborhood of 55 K can easily be maintained in this manner.

Observatory spacecraft instruments dedicated to sensing infrared or longer wavelengths need more elaborate cooling. The Hubble Space Telescope NICMOS (Near Infrared Camera and Multi-Object Spectrometer) originally used solid nitrogen coolant to reduce the detector temperature to 40 K, although it proved to be problematic. The Spitzer Infrared Telescope Facility (SIRTF) achieves a temperature of 1.4 K for its detectors by vapor cooling with helium effluent. BICEP, Background Imaging of Cosmic Extragalactic Polarization, is a ground-based instrument being developed to measure the polarization of the cosmic microwave background from Antarctica. Its detectors are cooled to 250 milli-Kelvens using a helium refrigeration system.

Scan Platforms

Optical instruments are sometimes installed on an articulated, powered appendage to the spacecraft bus called a scan platform, which points in commanded directions, allowing optical observations to be taken independently of the spacecraft's attitude. This is the case on Voyager and Galileo. Most newer spacecraft are designed without scan platforms, since a scan platform adds to a spacecraft's complement of mechanical and other subsystems, increasing mass and potential for failure. The alternative is to mount optical instruments directly to the spacecraft and rotate the entire spacecraft to point them. This approach is feasible due to the availability of large-capacity data storage systems that can record data while the spacecraft is off Earth-point carrying out observations.

Examples of Active Sensing Science Instruments

Synthetic Aperture Radar Imaging

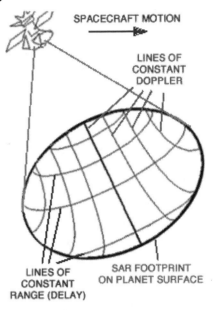

Some solar system objects that are candidates for radar imaging are covered by clouds or haze, making optical imaging difficult or impossible. These atmospheres are transparent to radio frequency waves, and can be imaged using Synthetic Aperture Radar (SAR) which provides its own penetrating illumination with radio waves. SAR is more a technique than a single instrument. It uses hardware and software, as most instruments do, but it also employs the motion of the spacecraft in orbit. SAR synthesizes the angular resolving power of an antenna many times the size of the antenna aperture actually used. SAR illuminates its target to the side of its direction of movement, and travels a distance in orbit while the reflected, phase-shift-coded pulses are returning and being collected. This motion during reception provides the basis for synthesizing an antenna (aperture) on the order of kilometers in size, using extensive computer processing.

Figure 12.10: SAR, Synthetic Aperture Radar imaging uses distance and Doppler to create a two-dimentional image.

For a SAR system to develop the resolution equivalent to optical images, the spacecraft's position and velocity must be known with great precision, and its attitude must be controlled tightly. This levies demands on the spacecraft's attitude control system, and requires spacecraft navigation data to be frequently updated. SAR images are constructed of a matrix where lines of constant distance or range intersect with lines of constant Doppler shift.

Magellan's radar instrument alternated its active operations as a SAR imaging system and radar altimeter with a passive microwave radiometer mode several times per second in orbit at Venus.

Examples of SAR imagery include Magellan's images of Venus,[7] and Cassini's images of Saturn's moon Titan.[8]

Altimeters

A spacecraft's altimeter sends coded radio pulses, or laser-light pulses, straight down to a planet's surface (the nadir) to measure variations in the height of the terrain below. The signals are timed from the instant they leave the instrument until they are reflected back, and the round-trip distance is obtained by dividing by the speed of light, and factoring in known equipment processing delays. Dividing by two then approximates the one-way distance between the instrument and the surface. Actual terrain height is then deduced based upon precise knowledge of the spacecraft's orbit.

Figure 12.11: MOLA on Mars Global Surveyor.

The Pioneer 12 spacecraft and the Magellan spacecraft used radar altimeters at Venus. Laser altimeters generally have a smaller footprint, and thus higher spatial resolution, than radar altimeters. They also require less power. The Mars Global Surveyor spacecraft carried a laser altimeter that uses a small cassegrain telescope. Known as MOLA for Mars Orbiter Laser Altimeter, its technology formed the basis for an experiment flown in Earth orbit in 1997 by the Space Shuttle.

Some Spacecraft Science Instrument Pages may be found here:

http://www.jpl.nasa.gov/basics/bsf12-1.php

Each of the following pages, linked from the above site, is a complete list of all of the science instruments on a spacecraft, with further links to information about each instrument. Among them you'll find direct- and remote-sensing instruments, and active and passive sensing instruments.

- Voyager Science Instruments
- Galileo Science Instruments
- Cassini Science Instruments
- Huygens Science Instruments
- Mars Global Surveyor Science Instruments
- Mars Express Science Instruments
- New Horizons Science Instruments
- Venus Express Science Instruments
- Messenger Science Instruments

- Ulysses Science Instruments
- Mars Exploration Rover (Spirit, Opportunity) Science Instruments
- Mars Pathfinder Science Instruments

Notes

[1] http://www-pw.physics.uiowa.edu/plasma-wave/voyager
[2] http://inms.gsfc.nasa.gov
[3] http://huygensgcms.gsfc.nasa.gov
[4] http://www.chemguide.co.uk/analysis/masspec/howitworks.html
[5] http://saturn.jpl.nasa.gov/faq/FAQRawImages
[6] http://pds-rings.seti.org/voyager/iss
[7] http://www.lpi.usra.edu/publications/slidesets/venus/
[8] http://pirlwww.lpl.arizona.edu/~perry/RADAR

Chapter 13

Spacecraft Navigation

Objectives: Upon completion of this chapter you will be able to describe basic ingredients of spacecraft navigation including spacecraft velocity and distance measurement, angular measurement, and how orbit determination is approached. You will be able to describe spacecraft trajectory correction maneuvers and orbit trim maneuvers. You will be able to recognize four distinct Deep Space Network data types used in navigation.

Spacecraft navigation comprises two aspects:

1. Knowledge and prediction of spacecraft position and velocity, which is orbit determination, and

2. Firing the rocket motor to alter the spacecraft's velocity, which is flight path control.

Recall from Chapter 4 that a spacecraft on its way to a distant planet is actually in orbit about the Sun, and the portion of its solar orbit between launch and destination is called the spacecraft's trajectory. Orbit determination involves finding the spacecraft's orbital elements (see page 84) and accounting for perturbations to its natural orbit. Flight path control involves commanding the spacecraft's propulsion system to alter the vehicle's velocity. Comparing the accurately determined spacecraft's trajectory with knowledge of the destination object's orbit is the basis for determining what velocity changes are needed.

Since the Earth's own orbital parameters and inherent motions are well known, the measurements we make of the spacecraft's motion as seen from Earth can be converted into the Sun-centered or heliocentric orbital parameters needed to describe the spacecraft's trajectory.

201

The meaningful measurements we can make from Earth of the spacecraft's motion are:

- Its distance or range from Earth,

- The component of its velocity that is directly toward or away from Earth, and

- To the extent discussed below, its position in Earth's sky.

Some spacecraft can generate a fourth type of nav data,

- Optical navigation, wherein the spacecraft uses its imaging instrument to view a target planet or body against the background stars.

By repeatedly acquiring these three or four types of data, a mathematical model may be constructed and maintained describing the history of a spacecraft's location in three-dimensional space over time. The navigation history of a spacecraft is incorporated not only in planning its future maneuvers, but also in reconstructing its observations of a planet or body it encounters. This is essential to constructing SAR (synthetic aperture radar) images, tracking the spacecraft's passage through planetary magnetospheres or rings, and interpreting imaging results.

Another use of navigation data is the creation of predicts, which are data sets predicting locations in the sky and radio frequencies for the Deep Space Network, DSN to use in acquiring and tracking the spacecraft.

Navigation Data Acquisition

The basic factors involved in acquiring the types of navigation data mentioned above are described below.

Spacecraft Velocity Measurement

Measurements of the Doppler shift of a coherent downlink carrier (see page 147) provide the radial component of a spacecraft's Earth-relative velocity. Doppler is a form of the tracking data type, TRK (see page 255), provided by the DSN.

Spacecraft Distance Measurement

A uniquely coded ranging pulse can be added to the uplink to a spacecraft and its transmission time recorded. When the spacecraft receives the ranging pulse, it returns the pulse on its downlink. The time it takes the spacecraft to turn the pulse around within its electronics is known from pre-launch testing. For example, Cassini takes 420 nanoseconds, give or take 9 ns. There are many other calibrated delays in the system, including the several microseconds needed to go from the computers to the antenna within DSN, which is calibrated prior to each use. When the pulse is received at the DSN, its true elapsed time at light-speed is determined, corrections are applied for known atmospheric effects (see page 104), and the spacecraft's distance is then computed. Ranging is also a type of TRK data provided by the DSN.

Distance may also be determined using angular measurement.

Spacecraft Angular Measurement

The spacecraft's position in the sky is expressed in the angular quantities Right Ascension and Declination (see page 51). While the angles at which the DSN antennas point are monitored with an accuracy of thousandths of a degree, they are not accurate enough to be used in determining a distant interplanetary spacecraft's position in the sky for navigation. DSN tracking antenna angles are useful only for pointing the antenna to the predicts given for acquiring the spacecraft's signal.

Fairly accurate determination of Right Ascension is a direct byproduct of measuring Doppler shift during a DSN pass of several hours. Declination can also be measured by the set of Doppler-shift data during a DSN pass, but to a lesser accuracy, especially when the Declination value is near zero, i.e., near the celestial equator. Better accuracy in measuring a distant spacecraft's angular position can be obtained using the following techniques:

Lateral Motion

No single measurement directly yields the lateral motion of a spacecraft deep in the solar system (if lateral motion means any component of motion except directly toward or away from Earth.)

But we do have a very good understanding of how things move in space – orbit models of spacecraft are built based on equations of motion from the likes of Kepler and Newton. There aren't many ways of moving (in other words, trajectories) that match up with a big batch of range and range-rate data acquired from various DSN station locations over a period or days, weeks or months.

The task is to apply measurements of Doppler and range to a model of a trajectory, and update that model to match all your measurements reasonably well, to obtain a solution to the orbit determination problem. Gaining knowledge of lateral motion is an iterative process.

VLBI: Extremely accurate angular measurements can be provided by a process independent from Doppler and range, VLBI, Very Long Baseline Interferometry. A VLBI observation of a spacecraft begins when two DSN stations on different continents (separated by a VLB) track a single spacecraft simultaneously. High-rate recordings are made of the downlink's wave fronts by each station, together with precise timing data. After a few minutes, both DSN antennas slew directly to the position of a quasar, which is an extragalactic object whose position on the plane of the sky is known to a high precision. Recordings are made of the quasar's radio-noise wavefronts.

Correlation and analysis of the recorded wavefronts yields a very precise triangulation from which the angular position may be determined by direct com- 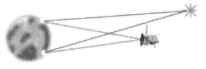 parison to the position of a quasar whose RA and DEC are well known. VLBI is considered a distinct DSN data type, different from TRK. This VLBI observation of a spacecraft is called a "delta DOR," DOR meaning differenced one-way ranging, and the "delta" meaning the difference between spacecraft and quasar positions.

Precision Ranging: Precision ranging refers to a set of procedures to ensure that range measurements are accurate to about 1 meter. Knowledge of  the spacecraft's Declination can be improved with Range measurements from two stations that have a large north-south displacement, for example between Spain and Australia, via triangulation.

Differenced Doppler: Differenced Doppler can provide a measure of a spacecraft's changing three-dimensional position. To visualize this, consider a spacecraft orbiting a distant planet. If the orbit is in a vertical plane exactly edge on to you at position A, you would observe the downlink to take a higher frequency as it travels towards you. As it recedes away from you to go behind the planet, you observe a lower frequency.

Now, imagine a second observer way across the Earth, at position B. Since the orbit plane is not exactly edge-on as that observer sees it, that person will record a slightly different Doppler signature. If you and the other observer were to compare notes and difference your data sets, you would have enough information to determine both the spacecraft's changing velocity and position in three-dimensional space. Two DSSs separated by a large baseline can do basically this. One DSS provides an uplink to the spacecraft so it can generate a coherent downlink, and then it receives two-way. The other DSS receives a three-way coherent downlink. The differenced data sets are frequently called "two-way minus three-way."

These techniques, combined with high-precision knowledge of DSN Station positions, a precise characterization of atmospheric refraction, and extremely stable frequency and timing references (F&T, which is another one of the DSN data types, see page 254), makes it possible for DSN to measure spacecraft velocities accurate to within hundredths of a millimeter per second, and angular position on the sky to within 10 nano-radians.

Optical Navigation

Spacecraft that are equipped with imaging instruments can use them to observe the spacecraft's destination planet or other body, such as a satellite, against a known background starfield. These images are called opnav images. The observations are carefully planned and uplinked far in advance as part of the command sequence development process. The primary body often appears overexposed in an opnav, so that the background stars will be clearly visible. When the opnav images are downlinked in telemetry (TLM, see page 255) they are immediately processed by the navigation team. Interpretation of opnavs provides a very precise data set useful for refining knowledge of a spacecraft's trajectory as it approaches a target. Note that this form of navigation data resides in the TLM data type.

Orbit Determination

The process of spacecraft orbit determination solves for a description of a spacecraft's orbit in terms of a state vector (position and velocity) at an epoch, based upon the types of observations and measurements described above. If the spacecraft is en route to a planet, the orbit is heliocentric; if it is in orbit about a planet, the orbit determination is made with respect to that planet. Orbit determination is an iterative process, building upon the results of previous solutions. Many different data inputs are selected as appropriate for input to computer software, which uses the laws of Newton. The inputs include the various types of navigation data described above, as well as data such as the mass of the Sun and planets, their ephemeris and barycentric movement, the effects of the solar wind and other non-gravitational effects, a detailed planetary gravity field model (for planetary orbits), attitude management thruster firings, atmospheric friction, and other factors.

Flight Path Control

An interplanetary spacecraft's course is mostly set once the launch vehicle has fallen away. From that point on, the spacecraft can make only very small corrections in its trajectory by firing small engines or thrusters. Often the largest complement of propellant that a spacecraft carries is reserved for orbit insertion at its destination.

Trajectory Correction Maneuvers

Once a spacecraft's solar or planetary orbital parameters are known, they may be compared to those desired by the project. To correct any discrepancy, a Trajectory Correction Maneuver (TCM) may be planned and executed. This adjustment involves computing the direction and magnitude of the vector required to correct to the desired trajectory. An opportune time is determined for making the change. For example, a smaller magnitude of change would be required immediately following a planetary flyby, than would be required after the spacecraft had flown an undesirable trajectory for many weeks or months. The spacecraft is commanded to rotate to the attitude in three-dimensional space computed for implementing the change, and its thrusters are fired for a determined amount of time. TCMs generally involve a velocity change (delta-V) on the order of meters, or sometimes tens of meters, per second. The velocity magnitude is necessarily small due to the limited amount of propellant typically carried.

Orbit Trim Maneuvers

Small changes in a spacecraft's orbit around a planet may be desired for the purpose of adjusting an instrument's field-of-view footprint, improving sensitivity of a gravity field survey, or preventing too much orbital decay.

Orbit Trim Maneuvers (OTMs) are carried out generally in the same manner as TCMs. To make a change increasing the altitude of periapsis, an OTM would be designed to increase the spacecraft's velocity when it is at apoapsis. To decrease the apoapsis altitude, an OTM would be executed at periapsis, reducing the spacecraft's velocity. Slight changes in the orbital plane's orientation may also be made with OTMs. Again, the magnitude is necessarily small due to the limited amount of propellant spacecraft typically carry.

Cassini provides an example of the accuracy achieved in rocket firings. The duration of a firing is executed within about 0.1% of the planned duration, and the pointing direction is executed within about 7 milliradians (0.4 degrees). Over the course of seven years from launch to arrival at Saturn, Cassini executed only seventeen of these planned, small velocity adjustments.

Historical Perspective

The process of orbit determination is fairly taken for granted today. During the effort to launch America's first artificial Earth satellites,[1] the JPL craft Explorers 1 and 2, a room-sized IBM computer was employed to figure a new satellite's trajectory using Doppler data acquired from Cape Canaveral and a few other tracking sites.

The late Caltech physics professor Richard Feynman[2] was asked to come to the Lab and assist with difficulties encountered in processing the data. He accomplished all of the calculations by hand, revealing the fact that Explorer 2 had failed to achieve orbit and had come down in the Atlantic ocean. The IBM mainframe was eventually coaxed to reach the same result, hours after Professor Feynman had departed for the weekend.

More on this story can be found in the book *Genius: The Life and Science of Richard Feynman* by James Gleick (Pantheon, 1992).

Notes

[1] http://www-spof.gsfc.nasa.gov/Education/wexp13.html
[2] http://www.feynmanonline.com

Part III

FLIGHT OPERATIONS

Chapter 14

Launch Phase

> **Objectives:** Upon completion of this chapter you will be able to describe the role launch sites play in total launch energy, state the characteristics of various launch vehicles, and list factors contributing to determination of launch periods and windows. You will be able to describe how the launch day of the year and hour of the day affect interplanetary launch energy and list the major factors involved in preparations for launch.

Mission Operations Phases

Interplanetary mission operations may be considered in four phases: the Launch Phase, the Cruise Phase, the Encounter Phase, and, depending on the state of spacecraft health and mission funding, the Extended Operations Phase. These "phases" are actually subcategories of a Flight Project's Phase E, Mission Operations and Data Analysis, as discussed in Chapter 7 (see page 109). Each of these mission operations phases, and subjects pertinent to them, are covered starting in this Chapter, and continuing in the next three Chapters.

Launch Vehicles

The launch of a spacecraft comprises a period of powered flight during which the vehicle rises above Earth's atmosphere and accelerates at least to orbital velocity. Powered flight ends when the rocket's last stage burns out, and the spacecraft separates and continues in freefall. If the spacecraft has achieved escape from Earth's gravitation, rather than entering Earth orbit, its flight path will then be purely a solar orbit of some description (since the launch pad was also in solar orbit).[1]

To date, the only practical way to produce the propulsive energy needed for launching spacecraft from Earth has been by combustion of chemical propellants. Mass drivers, however, may be useful in the future for launching material from the Moon or other small airless bodies. Ion-engine propulsion, whether powered by photovoltaics or nuclear reactors,[2] is useful not for launch phase, but for gently but steadily accelerating spacecraft that are already in Earth orbit or beyond.

Figure 14.1: The Centaur Upper Stage. Image courtesy NASA.

As first discussed in Chapter 3, there are two groups of propellants for chemical-combustion rockets, liquids and solids. Many spacecraft launches involve the use of both types of rockets, for example the solid rocket boosters attached to liquid-propelled rockets. Hybrid rockets, which use a combination of solid and liquid, are also being developed. Solid rockets are generally simpler than liquid, but they cannot be shut down once ignited. Liquid and hybrid engines may be shut down after ignition and, in some designs, can be re-ignited as needed.

Expendable launch vehicles, ELV, are used once. The U.S. Space Transportation System, STS, or Space Shuttle, was designed as a reuseable system to reach Low Earth Orbit. Most of its components are refurbished and reused multiple times.

Upper Stage Rockets can be selected for placement atop the launch vehicle's lower stages (or within the Shuttle's cargo bay) to provide the performance needed for a particular payload. The Centaur[3] high-energy upper stage has been a choice for robotic missions to the Moon and planets for many years. Figure 14.1 shows a full-scale Centaur rocket replica, configured with a model Surveyor spacecraft as the payload, Outside the NASA Glenn Visitor Center.

A sampling of launch vehicles of interest to interplanetary mission follows. One useful measure of performance for comparison among launch vehicles is the amount of mass it can lift to Geosynchronous Transfer Orbit, GTO (see page 85).

Delta

Delta is a family of two- or three-stage liquid-propelled ELVs that use multiple strap-on solid rocket boosters in several configurations. Originally made by McDonnell Douglas, it is now produced and launched by the Boeing and Lockheed Martin joint venture, United Launch Alliance (ULA).[4]

The Delta II, whose liquid-propellant engines burn kerosene and liquid oxygen (LOX), can be configured as two- or three-stage vehicles and with three, four or nine strap-on solid rocket graphite epoxy motors. Delta II payload delivery options range from about 891 to 2,142 kg to geosynchronous transfer orbit (GTO) and 2.7 to 6.0 metric tons to low-Earth orbit (LEO). Two-stage Delta II rockets typically fly LEO missions, while three-stage Delta II vehicles generally deliver payloads to GTO, or are used for deep-space explorations such as NASA's missions to Mars, a comet or near-Earth asteroids.

The Delta IV family of launch vehicles is capable of carrying payloads ranging from 4,210 kg to 13,130 kg to GTO. The three Delta IV Medium-Plus vehicles use a single common booster core that employs the RS-68 liquid hydrogen/liquid oxygen engine, which produces 2,949 kN (663,000 lb) of thrust. They are augmented by either two or four 1.5-meter diameter solid rocket strap-on graphite epoxy motors.

Figure 14.2: Delta II.[5]

Delta's launch record (350 flights as of early 2011) includes Earth orbiters and interplanetary missions dating back to 1960. Space Launch Complex 37, a historic Saturn-1 launch pad at the Kennedy Space Center (KSC), has become the site for the Delta-IV launch facility.

Titan

Titan was a family of U.S. expendable rockets used between 1959 and 2005. A total of 368 rockets of this family were launched. Titan IV, produced and launched for the U.S. Air Force by Lockheed Martin, was the nation's most powerful ELV until it was retired in 2005. Titan IV was capable of placing 18,000 kg into LEO, over 14,000 kg into polar orbit, or 4,500 kg into a geostationary transfer orbit (GTO). A Titan III launched the Viking spacecraft to Mars in 1975. A Titan IV, equipped with

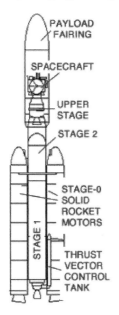

Figure 14.3: Titan IV.

two upgraded solid rocket boosters and a Centaur high-energy upper stage, launched the Cassini spacecraft on its gravity-assist trajectory to Saturn in 1997. Titan III vehicles launched JPL's Voyager 1 and 2 in 1977, and the Mars Observer spacecraft from the Kennedy Space Center (KSC), Cape Canaveral in 1992.

A Titan IV consisted of two solid-propellant stage-zero motors, a liquid propellant 2-stage core and a 16.7-ft diameter payload fairing. Upgraded 3-segment solid rocket motors increased the vehicle's payload capability by approximately 25%. The Titan IV configurations included a cryogenic Centaur upper stage, a solid-propellant Inertial Upper Stage (IUS), or no upper stage. Titan IV rockets were launched from Vandenberg Air Force Base, California, or Cape Canaveral Air Station, Florida.

Atlas

The Atlas V rocket is an expendable launch vehicle formerly built by Lockheed Martin. It is now built by the Lockheed Martin-Boeing joint venture United Launch Alliance. Aerojet develops and manufactures the Atlas V boosters. The rocket, built in Decatur, Alabama, consists of a first stage powered by kerosene and liquid oxygen, which uses a Russian made RD-180 engine, and a liquid hydrogen liquid oxygen powered Centaur upper stage. Some config-

Figure 14.4: Atlas V.[6]

urations also use strap-on booster rockets. Together these components are referred to as the Atlas launch vehicle. The Atlas-IIAS can put 3,833 kg in GTO. The Atlas-IIA was retired in December 2002, and the Atlas II was retired in March 1998. Re-engineering the Atlas-II booster resulted in an Atlas-III with at least a 5% improvement in performance.

The Atlas-V uses a "structurally stable" design; previous Atlas models depended on a pressurized propellant load for structural rigidity. With provisions for adding between one and five strap-on solid rocket boosters, the Atlas-V offers the capability of placing up to 8,670 kg in GTO. View a 1-minute movie of the first Atlas-V launch in August 2002 online.[7]

Space Launch Complex 41 at the Kennedy Space Center (KSC), where Viking-1 and -2, and Voyager-1 and -2 began their journeys, has become the site for the Atlas-V launch facility (see below).

Ariane

Today, Arianespace serves more than half the world's market for spacecraft launches to geostationary transfer orbit (GTO). Created as the first commercial space transportation company in 1980, Arianespace is responsible for the production, operation and marketing of the Ariane 5 launchers.[9]

Ariane 5, launched from the Kourou Space Center in French Guiana, entered service with a 6.5 metric ton payload capability to geostationary orbit. In 1996, the maiden flight of the Ariane 5 launcher ended in a failure. In 1997 Ariane 5's second test flight succeeded, and it is now in service.

Figure 14.5: Ariane V.[8]

Proton

The Proton is a liquid-propellant ELV, capable of placing 20,000 kg into LEO, originally developed by the Soviet CIS Interkosmos. It is launched from the Baykonur Kosmodrome in Kasakhstan, with launch services marketed by International Launch Services (a company formed in 1995 by Lockheed Martin, Khrunichev Enterprises and NPO Energia).[11] With an outstanding reliability record and over 200 launches, the Proton is the largest Russian launch vehicle in operational service. It is used as a three-stage vehicle primarily to launch large space station type payloads into low Earth orbit, and in a four-stage configuration to launch spacecraft into GTO and interplanetary trajectories.

Figure 14.6: Proton.[10]

Soyuz

The proven Soyuz launch vehicle is one of the world's most reliable and frequently used launch vehicles. As of early 2011, more than 1,766 missions have been performed by Soyuz launchers to orbit satellites for telecommunications, Earth observation, weather and scientific missions, as well as for piloted flights. Soyuz

Figure 14.7: Soyuz.[12]

is being evolved to meet commercial market needs, offering payload lift capability of 4,100 kg. to 5,500 kg. into a 450-km. circular orbit. Souyz is marketed commercially by Starsem, a French-registered company.[13]

Space Transportation System

America's Space Shuttle, as the Space Transportation System (STS) is commonly known,[15] is a reusable launching system whose main engines burn liquid hydrogen and LOX. After each flight, its main components, except the external propellant tank, are refurbished to be used on future flights. The STS can put payloads of up to 30,000 kg in LEO. With the appropriate upper stage, spacecraft may be boosted to a geosynchronous orbit or injected into an interplanetary trajectory. Galileo, Magellan, and Ulysses were launched by the STS, using an Inertial Upper Stage (IUS), which is a two-stage solid-propellant vehicle. The STS may

Figure 14.8: Space Shuttle Atlantis.[14]

be operated to transport spacecraft to orbit, perform satellite rescue, assemble and service the International Space Station, and to carry out a wide variety of scientific missions ranging from the use of orbiting laboratories to small self-contained experiments.

Smaller Launch Vehicles

Many NASA experiments, as well as commercial and military payloads, are becoming smaller and lower in mass, as the art of miniaturization advances. The range of payload mass broadly from 100 to 1300 kg is becoming increasingly significant as smaller spacecraft are designed to have more op-

Figure 14.9: Pegasus.[16]

erational capability. The market for launch vehicles with capacities in this range is growing.

Pegasus is a small, winged solid-propellant ELV built and flown by Orbital Sciences Corporation.[17] It resembles a cruise missile, and is launched from under the fuselage of an aircraft while in flight at high altitude, currently Orbital Sciences' L-1011. Pegasus can lift 400 kg into LEO.

Taurus is the ground-based variant of Orbital Sciences Corporation's air-launched Pegasus rocket. This four-stage, transportable, inertially guided, all solid propellant vehicle is capable of putting 1,350 kg into LEO, or 350 kg in GTO.

Saturn V

The Saturn V launch vehicle served admirably during the Apollo Program that landed a dozen humans on the Moon. Although this vehicle was never used again, honorable mention seems appropriate. This most powerful rocket ever launched was developed at NASA's Marshall Space Flight Center under the direction of Wernher von Braun. Fifteen were built. Its fueled first stage alone weighed more than an entire Space Shuttle.

Launch Sites

If a spacecraft is launched from a site near Earth's equator, it can take optimum advantage of the Earth's substantial rotational speed. Sitting on the launch pad near the equator, it is already moving at a speed of over 1650 km per hour relative to Earth's center. This can be applied to the speed required to orbit the Earth (approximately 28,000 km per hour). Compared to a launch far from the equator, the equator-launched vehicle would need less propellant, or a given vehicle can launch a more massive spacecraft.

Figure 14.10: Launch Complex 41 at Kennedy Space Center, with an Atlas-V poised for launch amid four lightning rod towers (LC-41 was previously the launch pad for the Viking & Voyager missions). Image courtesy NASA.

A spacecraft intended for a high-inclination Earth orbit has no such free ride, though. The launch vehicle must provide a much larger part, or all, of the energy for the spacecraft's orbital speed, depending on the inclination.

For interplanetary launches, the vehicle will have to take advantage of Earth's orbital motion as well, to accommodate the limited energy available from today's launch vehicles. In the diagram below, the launch vehicle is accelerating generally in the direction of the Earth's orbital motion (in addition to using Earth's rotational speed), which has an average velocity of approximately 100,000 km per hour along its orbital path.

In the case of a spacecraft embarking on a Hohmann interplanetary transfer orbit (see page 75), recall the Earth's orbital speed represents the speed at aphelion or perihelion of the transfer orbit, and the spacecraft's velocity merely needs to be increased or decreased in the tangential direction to achieve the desired transfer orbit.

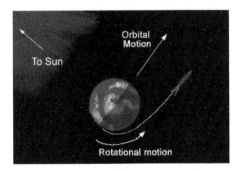

Figure 14.11: Making use of Earth's motion.

The launch site must also have a clear pathway downrange so the launch vehicle will not fly over populated areas, in case of accidents. The STS has the additional constraint of requiring a landing strip with acceptable wind, weather, and lighting conditions near the launch site as well as at landing sites across the Atlantic Ocean, in case an emergency landing must be attempted.

Launches from the east coast of the United States (the Kennedy Space Center at Cape Canaveral, Florida) are suitable only for low inclination orbits because major population centers underlie the trajectory required for high-inclination launches. High-inclination launches are accomplished from Vandenberg Air Force Base on the west coast, in California, where the trajectory for high-inclination orbits avoids population centers. An equatorial site is not preferred for high-inclination orbital launches. They can depart from any latitude.

Complex ground facilities are required for heavy launch vehicles, but smaller vehicles such as the Taurus can use simpler, transportable facilities. The Pegasus requires none once its parent airplane is in flight.

Launch Period

A launch period is the span of time during which a launch may take place while satisfying the constraints imposed by safety and mission objectives. For an interplanetary launch, the window is constrained typically within a number of weeks by the location of Earth in its orbit around the Sun, in order to permit the vehicle to use Earth's orbital motion for its trajectory, while timing it to arrive at its destination when the target planet is in position.

A launch *window* may also be specified, constrained to a number of hours or minutes each day during the launch period, in order to take best advantage of Earth's rotational motion. In the illustration above, the vehicle is launching from a site near the Earth's terminator which is going into night as the Earth's rotation takes it around away from the Sun. If the example in the illustration were to launch in the early morning hours on the other side of the depicted Earth, it would be launching in a direction opposite Earth's orbital motion.

These examples are over-simplified in that they do not differentiate between launch from Earth's surface and injection into interplanetary trajectory. It is actually the latter that must be timed to occur on the proper side of Earth. Actual launch times must also consider how long the spacecraft needs to remain in low Earth orbit before its upper stage places it on the desired trajectory (this is not shown in the illustration).

The daily launch window may be further constrained by other factors, for example, the STS's emergency landing site constraints. Of course, a launch which is to rendezvous with another vehicle in Earth orbit must time its launch with the orbital motion of that object. This has been the case with the Hubble Space Telescope repair missions executed in December 1993, February 1997, December 1999, March 2002, and May 2009.

Preparations For Launch

ATLO stands for Assembly, Test, and Launch Operations. This period is usually scheduled very tightly. Spacecraft engineering components and instruments are all delivered according to plan where the spacecraft first takes shape in a large clean room. They are integrated and tested using computer programs for command and telemetry very much like those that will be used in flight. Communications are maintained with the growing spacecraft nearly continuously throughout ATLO. The spacecraft is transported to an environmental test lab where it is installed on a shaker table and subjected to launch-like vibrations. It is installed in a thermal-vacuum chamber to test its thermal properties, all the while communicating with engineers. Adjustments are made as needed in thermal blanketing, and thermal-vacuum tests may be repeated.

Then the spacecraft is transported to the launch site. The spacecraft is sealed inside an environmentally controlled carrier for the trip, and internal conditions are carefully monitored throughout the journey whether it is by truck or airplane.

Once at the launch site, additional testing takes place. Propellants are loaded aboard. Any pyrotechnic devices are armed. Then the spacecraft is mated to its upper stage, and the stack is hoisted and mated atop the launch vehicle. Clean-room conditions are maintained atop the launch vehicle while the payload shroud (fairing) is put in place.

Figure 14.12: Cassini-Huygens Spacecraft being mated to its launch vehicle adapter.

Pre-launch and launch operations of a JPL spacecraft are typically carried out by personnel at the launch site while in direct communication with persons at the Space Flight Operations Facility at JPL. Additional controllers and engineers at a different location may be involved with the particular upper stage vehicle, for example the Lockheed personnel at Sunnyvale, California, monitoring performance of the inertial upper stage (IUS) or the Centaur upper stage. The spacecraft's telecommunications link is maintained through ground facilities close to the launch pad prior to launch and during launch, linking the spacecraft's telemetry to controllers and engineers at JPL. Command sequences must be loaded aboard the spacecraft, verified, and initiated at the proper time prior to launch. Spacecraft health must be monitored, and the launch process interrupted if any critical tolerances are exceeded.

As soon as the spacecraft is launched, the DSN begins tracking, acquiring the task from the launch-site tracking station, and the mission's cruise phase is set to begin.

Perspective: A Launch to Mars

Contributed by JPL's Arden Acord, who served as the Mars Reconnaissance
Orbiter (MRO)[18] Launch Vehicle Liaison for launch in August 2005.

Buying Launch

The whole process, from buying the launch service through the actual launch,
took about four and a half years for MRO. For most missions that we fly,
we don't actually buy a launch vehicle, we buy a "launch service." Kennedy
Space Center (KSC) manages the acquisition process with the support of
project personnel. KSC basically has a catalog of launch vehicles that are
pre-qualified to fly our class of spacecraft, along with a "not-to-exceed"
price for each one. Since there may be more than one rocket type that can
perform a mission, KSC may conduct a bid process for each of the launch
service providers (e.g., Boeing, Lockheed Martin) to compete for our business. Project people are involved to make sure that all the requirements
are right. Once the contract is awarded by NASA, we start the integration
process.

Launch Vehicle Integration

There is a long, detailed process of planning how we integrate the spacecraft
to the launch vehicle so that everything will work properly. This includes
making sure the spacecraft will fit inside the launch vehicle's payload fairing,
making sure the spacecraft won't be damaged by the forces put on it during
launch, and ensuring that the launch vehicle will put the spacecraft on the
right trajectory to get where it needs to go.

The spacecraft is shipped to the launch site at about the same time
the launch vehicle is shipped there, three months or so before launch. These
two systems go through their final preparations for physical integration. The
spacecraft is attached to the launch vehicle adapter, and is then encapsulated
in the fairing. The encapsulated spacecraft is transported to the launch pad
and is hoisted by crane and attached to the top of the launch vehicle. This

Figure 14.13: Arden signals "thumbs up" as MRO is hoisted atop the Atlas V.

happens about ten days before launch, at which time when the countdown clock typically starts.

The countdown helps everyone orchestrate the many parallel operations that are going on to get everything ready. People have to work at very specific times during the countdown so they don't interfere with one other. During this period, any final operations on the spacecraft take place, including removal of instrument covers and other "remove before flight" items, installation of arming plugs, and generally getting everything buttoned up for the big day.

Launch Day

Around one day before launch, things start to get exciting. The launch vehicle begins its fueling process and final preparations. The spacecraft launch crews come in to power up the spacecraft, load software and send commands that put the spacecraft into its launch configuration. At various times, there are "polls" where the launch managers (including the project manager) report whether they are "go for launch." About five or ten minutes before launch, the spacecraft is in its launch state, and the launch vehicle goes through its final countdown. During these final minutes, the launch process is controlled mostly by computers. At "T minus zero," if there are no aborts, the rocket with our spacecraft on top begins its journey of discovery.

Notes

[1] http://www.braeunig.us/space/orbmech.htm#launch
[2] http://trajectory.grc.nasa.gov/projects/ntp
[3] http://www.nasa.gov/centers/glenn/about/history/centaur.html
[4] http://www.ulalaunch.com
[5] Illustration courtesy Boeing.
[6] Image courtesy United Launch Alliance.
[7] http://www.ilslaunch.com/newsroom/video-gallery
[8] Image courtesy Arianespace.
[9] http://www.arianespace.com
[10] Image courtesy International Launch Services.
[11] http://www.ilslaunch.com
[12] Image courtesy Starsem.
[13] http://www.starsem.com
[14] Image courtesy NASA.
[15] http://www.nasa.gov/mission_pages/shuttle
[16] Image courtesy Orbital Sciences Corporation.
[17] http://www.orbital.com
[18] http://mpfwww.jpl.nasa.gov/mro

Chapter 15

Cruise Phase

> **Objectives:** Upon completion of this chapter, you will be able to list the major factors involved in a mission's cruise phase, including spacecraft checkout and characterization, and preparation for encounter. You will be able to characterize typical daily flight operations.

Cruise phase is bounded by launch phase at the beginning and encounter phase at the end. It may be as short as a few months, or it may span years with the spacecraft looping the Sun to perform gravity-assist planetary flybys. It is a time during which ground system upgrades and tests may be conducted, and spacecraft flight software modifications are implemented and tested. Cruise

Figure 15.1: Messenger.[1]

operations for JPL missions are typically carried out from the Space Flight Operations Facility at JPL.

As of this printing in early 2011, the APL[2] Messenger spacecraft[3] has ended its 6.5-year cruise phase. Its gravity-assist flybys included one of Earth, two of Venus, and three of Mercury, all in the process of shedding momentum in solar orbit. It entered orbit around Mercury on March 18, 2011 to study the planet for at least one Earth year. This mission will obtain the first new data from Mercury in more than 30 years.

Spacecraft Checkout and Characterization

After launch, the spacecraft is commanded to configure for cruise. Appendages that might have been stowed in order to fit within the launch vehicle are deployed either fully or to intermediate cruise positions. Telemetry

225

is analyzed to determine the health of the spacecraft, indicating how well it survived its launch. Any components that appear questionable might be put through tests specially designed and commanded in or near real time, and their actual state determined as closely as possible by subsequent telemetry analysis.

During the cruise period, additional command sequences are uplinked and loaded aboard for execution, to replace the command sequence exercised during launch. These take the spacecraft through its routine cruise operations, such as tracking Earth with its HGA and monitoring celestial references for attitude control. Flight team members begin to get the feel of their spacecraft in flight.

Commonly, unforeseen problems arise, and the onboard fault protection algorithms receive their inevitable tests; the spacecraft will, more likely than not, go into safing or contingency modes (as described in Chapter 11), and nominal cruise operations must be painstakingly recovered.

TCMs (Trajectory Correction Maneuvers), are executed to fine-tune the trajectory. As the spacecraft nears its target, or earlier during designated checkout periods, the science instruments are powered on, exercised and calibrated.

Real-time Commanding

Frequently, command sequences stored on the spacecraft during cruise or other phases must be augmented by commands sent and executed in or near real time, as new activities become desirable, or, less frequently, as mistakes are discovered in the on-board command sequence.

There are risks inherent in real-time commanding. It is possible to select the wrong command file for uplink, especially if an extensive set of files exists containing some similar or anagrammatic filenames. Also, it is possible that a command file built in haste may contain an error. Planned sequences of commands (generally just called "sequences") are typically less risky than real-time commands because they benefit from a long process of extensive debate and selection, testing and checking and simulation prior to uplink.

Some flight projects permit real-time commanding of a sort wherein command mnemonics are entered at the command system keyboard in real time and uplinked directly. Other projects do not permit this inherently high-risk operation because even a typographical error passing undetected might introduce a problem with the spacecraft or the mission.

These factors may limit the desirability of undertaking many activities by real-time commands, but the necessity, as well as the convenience, of at least some form of real-time commanding frequently prevails.

Typical Daily Operations

Usually, at least one person is on duty at the Flight Project's realtime mission support area at JPL or other location, during periods when the DSN (Deep Space Network),[4] is tracking the spacecraft. This person, typically the mission controller, or "Ace," watches realtime data from the

Figure 15.2: Cassini Aces during early cruise.

spacecraft and ground system, and responds to any anomalous indications via pre-planned responses. Anomaly response plans usually include getting in touch with appropriate subsystem experts in real time, no matter what time it is locally, to evaluate the situation and determine how best to proceed.

The Ace is a person on the Mission Control Team or Real-time Operations Team who is the single point of contact between the entire flight team, consisting of, for example, a Spacecraft Team, a Navigation Team, Science Teams, and other teams on the one hand, and teams external to the flight project such as DSN, Facilities Maintenance, multimission Data Systems Operations Team (DSOT), Ground Communications Facility (GCF), Advanced Multi-Mission Operations System (AMMOS), System Administrators (SAs), Network Administrators, and others, on the other hand.

"Ace" is not an acronym, despite attempts to make it one. It simply refers to one single point of contact for a project's real-time flight operations, and is not too inappropriately a pun for an expert pilot. Most Aces have experience with many different missions. It is possible for one Ace to be serving more than one flight project at a time. In 1993, one particular Ace (who is depicted in the banner at the top of this page) was serving Magellan, Voyager 1, Voyager 2, and Mars Observer, all at the same time during regular work shifts.

The Ace executes commanding, manages the ground systems, insures the capture and delivery of telemetry, monitor, and tracking data, watches for alarm limit violations in telemetry, manages the alarm limits, evaluates data quality, performs limited real-time analyses to determine such things as maneuver effectiveness and spacecraft health to first order, and coördinates the activities of the DSN and other external teams in support of the flight project(s). Typically, a large portion of the Ace's interactions are with the Spacecraft Team, the DSN, and the DSOT. On the infrequent occasions when the Ace detects an anomaly and rules out any false indications, he or she proceeds to invoke and follow the anomaly response plan that has been approved by the flight project. That plan is then followed by appropriate flight team members until the anomaly is resolved, and nominal operations are restored.

Monitoring Spacecraft and Ground Events

In order to tell whether everything is proceeding nominally, an Ace needs an accurate list of expected events, to compare with spacecraft events as they are observed in real time. For example, the spacecraft's downlink signal may change or disappear. Was it planned, or is this an anomaly?

Such a list is also required for the purpose of directing DSN station activity, and for planning command uplinks and other real-time operations. For example, if we uplink a command at 0200 UTC, will the spacecraft actually receive it one-way light-time later, or will the spacecraft be off Earthpoint or behind a planet at that time?

That list is called the sequence of events, SOE. It contains a list of spacecraft events being commanded from the onboard sequence, and DSN ground events such as the beginning and end of tracking periods, transmitter operations, and one-way light times. See page 59 for a sample SOE page for reference, and a discussion of the various time conventions used in the SOE.

Compiling an SOE and related products begins with a list of the commands that will be uplinked to the spacecraft's sequencing memory, and that will execute over a period of typically a week or a month or more into the future. The list of commands, sorted into time-order, comes from engineers responsible for spacecraft subsystems, scientists responsible for their instruments' operations, and from others. Times for the events' execution are included with the commands. The team responsible for generating the command sequence then creates a spacecraft event file (SEF). This file goes on as an input to the remainder of the sequence generation process for eventual uplink to the spacecraft.

A copy of the SEF also goes to the sequence of events generator software, SEGS, where commands are adjusted for light time, and are merged with DSN station schedule information and events. Station viewperiod files and light time files are typically provided by the navigation team. One of

SEGS output products is the DSN keyword file (DKF). This file is provided to the DSN, who then combines it with similar listings from other projects to create an SOE for each particular station. SEGS outputs the SOE in tabular form, and also arranges most of the same information into a high-level graphics product, the space flight operations schedule, SFOS. Users can view each of these products, or create hardcopy, using SEGS viewing and editing software. The illustration below shows activities typical for the generation of SOE products.

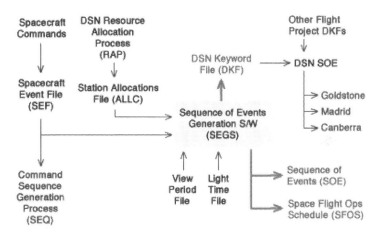

Figure 15.3: Creating products for use in flight operations.

Tracking the Spacecraft in Flight

DSN tracking requirements and schedules have been negotiated months or even years in advance of launch. Now the spacecraft is in flight. Near the time when the spacecraft will be rising in the sky due to Earth's rotation, its assigned DSN tracking activity begins.

Precal

During the period allotted for "precalibration" activities, the Network Monitor and Control (NMC) operator sits down at her or his console in the Signal Processing Center (SPC) of one the DSN's three Deep Space Communications Complexes (DSCC). The operator will be controlling and monitoring the assigned antenna, called a Deep Space Station (DSS), as well as an assigned set of computers that control its pointing, tracking, commanding, receiving, telemetry processing, ground communications, and other functions.

This string of equipment from the antenna to the project people at JPL is called a "connection", referring to the two-way communications link between the spacecraft and the project. The NMC operator in this role is called the "Connection Controller." Prior to the Connection Controller's arrival, the Complex Manager (a higher-level NMC function) operator will have assigned, via directives sent out to the station components over a local area network (LAN), applicable equipment to become part of the connection.

Figure 15.4: Deep Space Station 14 pointing its 70 m aperture toward the eastern horizon in the Mojave Desert.

At about this time, a JPL Communications technician establishes a voice link between the Connection Controller and the Ace. The Ace offers a pre-pass briefing to the Connection Controller, highlighting the important activities for the upcoming pass, and any changes in the latest documentation. Now the Connection Controller begins sending more directives over the SPC LAN to configure each of the link components specifically for the upcoming support. Predict sets containing uplink and downlink frequencies, Doppler bias ramp rates, pointing angles and bit rates, command modulation levels, and hundreds of other parameters, are all sent to the link components. Any problems are identified and corrected as the Connection Controller and the Ace communicate as needed during the "precal."

Near the end of the precal period, the Connection Controller checks the DSS area via closed circuit TV, makes a warning announcement over its outdoor loudspeakers, and the DSS antenna swings to point precisely to the spacecraft's apparent location (see page 104) above the eastern horizon.

BOT

Beginning of Track (BOT) is declared. At the prearranged time, typically ten minutes after BOT, the DSS's transmitter comes on, and red rotating beacons on the antenna illuminate as a warning of the microwave power present. Upon locking the receivers, telemetry, and tracking equipment to the spacecraft's signal, the link is established. This marks the Beginning of Track (BOT) and Acquisition of Signal (AOS).

The Ace interacts with the Connection Controller as needed during the course of the track to be sure the flight project's objectives are met. Depending on the nature of the spacecraft's activities, there may be Loss of Signal (LOS) temporarily when the spacecraft turns away to maneuver, or if it goes into occultation behind a planet. This LOS would presumably be followed by another AOS when the maneuver or occultation is complete. During the day, the DSS antenna moves slowly to follow, or track, the spacecraft as the Earth rotates.

EOT

Near the end of the Connection Controller's shift, the DSS is pointing lower on the western horizon. At the same time, if continuous tracking is scheduled, another Connection Controller inside the SPC of another DSCC a third of the way around the world, is in touch with the Ace, conducting a precal as the same spacecraft is rising in the east from the new station's point of view. To accomplish an uplink transfer, the setting DSS's transmitter is turned off precisely two seconds after the rising DSS's transmitter comes on. Scheduled End of Track (EOT) arrives, and the Connection Controller at the setting DSS begins postcal activities, idling the link components and returning control of them to the Complex Manager operator. After a debriefing with the Ace, the Ace releases the station, so it can begin preparing for its next scheduled activity.

Preparation for Encounter

Command loads uplinked to the spacecraft are valid for varying lengths of time. So-called quiescent periods such as the lengthy cruises between planets require relatively few activities, and a command load may be valid for several weeks or even months. By comparison, during the closest-approach part of a flyby encounter of a prime target body, a very long and complex load may execute in only a matter of hours.

Prior to the Voyagers' encounters, the spacecraft was generally sent a command sequence that took it through activities simulating the activities of encounter. Nowadays, this kind of rehearsal is largely accomplished with ground-based simulation, using simulated data broadcast to users. Changes in data rate and format, and spacecraft maneuvers, are designed to put the flight team and ground systems through their paces during a realistic simulation, in order to provide some practice for readiness, to shake down the systems and procedures, and to try to uncover flaws or areas for improvement. In a few critical cases, the actual encounter command load, slightly modified so that some activities (such as rocket firings) are skipped, may be actually executed aboard the spacecraft as a final test. This was the case with Cassini's Saturn Orbit Insertion critical sequence, and also with its Huygens Mission Relay critical sequence.

Instrument calibrations are undertaken prior to encounter, and again afterwards, to be sure that experiments are being carried out in a scientifically controlled fashion.

Notes

[1]Image courtesy Johns Hopkins University Applied Physics Laboratory.
[2]http://www.jhuapl.edu
[3]http://messenger.jhuapl.edu
[4]http://deepspace.jpl.nasa.gov

Chapter 16

Encounter Phase

> **Objectives:** Upon completing this chapter, you will be able to describe major factors involved in flyby operations, planetary orbit insertion, planetary system exploration, planet mapping, and gravity field surveying. You will be able to describe the unique opportunities for science data acquisition presented by occultations, and problems involved. You will be able to describe the concepts of using aerobraking to alter orbital geometry or decelerate for atmospheric entry, descent and landing.

The term "encounter" is used in this chapter to indicate the high-priority data-gathering period of operations for which the mission was intended. It may last a few months or weeks or less as in the case of a flyby encounter or atmospheric probe entry, or it may last a number of years as in the case of an orbiter. For JPL missions, brief critical portions of encounter operations are typically carried out from a special room within the Space Flight Operations Facility that is equipped for television cameras and VIP visits.

Flyby Operations

All the interplanetary navigation and course corrections accomplished during cruise result in placement of the spacecraft at precisely the correct point, and at the correct time to carry out its encounter observations. A flyby spacecraft has a limited opportunity to gather data. Once it has flown by its target, it cannot return to recover any lost data. Its operations are planned years in advance of the encounter, and the plans are refined and practiced in the months prior to the encounter date. Sequences of commands are prepared, tested, and uplinked by the flight team to carry out operations in various phases of the flyby, depending on the spacecraft's distance from its target.

During each of the six Voyager encounters, the phases were called Observatory phase, Far Encounter phase, Near Encounter phase, and Post Encounter phase. These phases may be given other names on different missions, but many of the functions most likely will be similar. Many of these also apply in some degree to missions other than flyby missions as well. In fact, for a spacecraft orbiting Jupiter or Saturn, for example, flyby encounters occur repeatedly, as the spacecraft approaches each targeted satellite, on orbit after orbit.

In a flyby operation, Observatory phase (OB) begins when the target can be better resolved in the spacecraft's optical instruments than it can from Earth-based instruments. This phase generally begins a few months prior to the date of flyby. OB is marked by the spacecraft being for the first time completely involved in making observations of its target (rather than cruise activities), and ground resources are completely operational in support of the encounter. The start of this phase marks the end of the interplanetary cruise phase. Ground system upgrades and tests have been

Figure 16.1: Planetary flyby.

completed, spacecraft flight software modifications have been implemented and tested, and the encounter command sequences have been placed on board.

Far Encounter phase (FE) begins when the full disc of a planet can no longer fit within the field of view of the instruments. Observations are designed to accommodate parts of the planet rather than the whole disc, and to take best advantage of the higher resolution available.

Near Encounter phase (NE) includes the period of closest approach to the target. It is marked by intensely active observations with all of the spacecraft's science experiments, including onboard instruments and radio science investigations. It includes the opportunity to obtain the highest resolution data about the target. During NE, radio science observations may include ring plane measurements during which ring structure and particle sizes can be determined. Celestial mechanics observations can determine the planet's or satellites' mass, and atmospheric occultations can determine atmospheric structures and compositions.

During the end of FE or the beginning of NE, a bow shock crossing may be identified through data from the magnetometer, the plasma instrument and plasma wave instrument as the spacecraft flies into a planet's magnetosphere and leaves the solar wind. When the solar wind is in a state of flux, these crossings may occur again and again as the magnetosphere and the solar wind push back and forth over millions of kilometers.

Post encounter phase (PE) begins when NE completes, and the spacecraft is receding from the planet. It is characterized by day after day of observations of a diminishing, thin crescent of the planet just encountered. This is the opportunity to make extensive observations of the night side of the planet. After PE is over, the spacecraft stops observing its target planet, and returns to the activities of cruise phase. DSN resources are relieved of their continuous support of the encounter, and they are generally scheduled to provide less frequent coverage to the mission during PE.

After encounter, instrument calibrations are repeated to be sure that any changes in the instruments' states are accounted for.

Late Updates

While observations must be planned in detail many months or years prior to NE, precise navigation data may not be available to command accurate pointing of the instruments until only a few days before the observations execute. So, late updates to stored parameters on the spacecraft can be made to supply the pointing data just in time. OPNAVs, discussed in Chapter 13, may be an important navigational input to the process of determining values for late parameter updates.

Also, some observations of the target planet, or its satellites or environs may be treated as reprogrammable late in the encounter, in order to observe features that had not been seen until FE. Commands to cover new targets might be uplinked, or an entire block of observations may be commanded to shift in time.

Planetary Orbit Insertion Operations

The same type of highly precise interplanetary navigation and course correction used for flyby missions also applies during cruise for an orbiter spacecraft. This process places the spacecraft at precisely the correct location at the correct time to enter into planetary orbit. Orbit insertion requires not only precise position and timing, but also controlled deceleration. As the spacecraft's trajectory is bent by the planet's gravity, the command sequence aboard the spacecraft places the spacecraft in the correct attitude, and fires its engine(s) at the proper moment and for the proper duration. Once the retro-burn has completed, the spacecraft has been captured into orbit by its target planet. If the retro-burn were to fail, the spacecraft would continue to fly on past the planet as though it were a flyby mission. It is common for the retro-burn to occur on the far side of a planet as viewed from Earth, with little or no data available until well after the burn has completed and the spacecraft has emerged from behind the planet, successfully in orbit.

Once inserted into a highly elliptical orbit, Mars Global Surveyor continued to adjust its orbit via aerobraking (discussed later in this chapter) near periapsis to decelerate the spacecraft further, causing a reduction in the apoapsis altitude, and establishing a close circular orbit at Mars. Mars Odyssey used the same technique. Galileo used a gravity assist from a close flyby of Jupiter's moon Io to decelerate, augmenting the deceleration provided by its 400 N rocket engine. Thereafter, additional Orbit Trim Maneuvers (OTMs) over a span of several years were used to vary the orbit slightly and choreograph multiple encounters with the

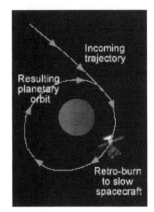

Figure 16.2: Orbit insertion.

Galilean satellites and the magnetosphere. Cassini is currently doing the same in orbit at Saturn.

System Exploration or Planetary Mapping

At least two broad categories of orbital science-gathering operations may be identified: system exploration and planetary mapping. Exploring a planetary system includes making observations of the planet, its atmosphere, its satellites, its rings, and its magnetosphere during a tour typically a few years or more in duration, using the spacecraft's complement of remote-sensing and direct-sensing instruments. On the other hand, mapping a planet means concentrating observations on the planet itself, using the spacecraft's instruments to obtain data mainly from the planet's surface and atmosphere.

Galileo explored the entire Jovian system, including its satellites, rings, magnetosphere, the planet, its atmosphere, and its radiation environment. At Saturn, Cassini is engaged in a similar exploratory mission, examining the planet's atmosphere, rings, magnetosphere, icy satellites, and the large satellite Titan, which has its own atmosphere. Magellan, a planetary mapper, covered more than 99% the surface of Venus in great detail using Synthetic Aperture Radar (SAR) imaging, altimetry, radiometry, gravity field, and mass distribution. A few special experiments were also carried out, including bistatic radar, aerobraking, windmilling, and destructive atmospheric entry. Mars Global Surveyor and Mars Odyssey are mapping the surface of their planet using imaging, altimetry, spectroscopy, and a gravity field survey.

An orbit of low inclination at the target planet (equatorial, for example) is well suited to a system exploration mission, because it provides repeated exposure to satellites orbiting within the equatorial plane, as well as adequate coverage of the planet and its magnetosphere. An orbit of high inclination (polar, for example) is better suited for a mapping mission, since the target planet or body will rotate fully below the spacecraft's orbit, providing eventual exposure to every part of the planet's surface.

In either case, during system exploration or planetary mapping, the orbiting spacecraft is involved in an extended encounter period, requiring continuous or dependably regular support from the flight team members, the DSN, and other institutional teams.

Occultations

Occultations provide unique opportunities to conduct scientific experiments. Occultations of interest include the Earth, the Sun, or another star disappearing behind a planet, behind its rings, or behind its atmosphere, as viewed from the spacecraft. They are of interest when involving either the main planet or its satellites. During the one-time only opportunity for occultation by a planet during a flyby mission encounter, or repeatedly during an orbital mission, onboard instruments may make unique observations. Here are a few examples:

- An ultraviolet spectrometer watches the Sun as it disappears behind a planet's atmosphere (or a satellite's atmosphere), obtaining spectra deep into the atmosphere, that can be studied to determine its composition and structure.

- A photometer watches a bright distant star as it passes behind a ring system. Results yield high-resolution data on the sizes and structures of the ring system and its particles.

- The telecommunications system is used by Radio Science to measure the Doppler shift around closest approach to a satellite, to determine the object's mass.

- Radar Science commandeers the telecommunications system to perform a bistatic radar experiment: The Deep Space Network (DSN) observes the radio signal glancing off a satellite's surface as the spacecraft directs its downlink signal at just the right spots on the satellite to achieve this specular reflection. Results can yield information on the structure and composition of the surface. (While this kind of experiment doesn't actually need an occultation, it might be performed near occultation.)

- Radio Science observes the spacecraft's radio signal on Earth as the spacecraft passes behind a planet or a satellite, yielding data on the composition and structure of its atmosphere (see Chapter 8).

- Radio Science ring occultations provide data on the ring system's structure and composition. Of course this involves recording the downlink as the DSN sees it through the rings. The spacecraft's transmission is trained on Earth as the rings drift by.

Recall that the most precise Radio Science investigations require a two- or three-way coherent mode, receiving an uplink from the DSN as discussed in Chapter 10. However, with an atmospheric occultation, or an occultation of an opaque ring section, this coherent link is generally possible on ingress only; the spacecraft is likely to lose the uplink from DSN when it

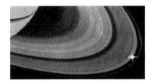

Figure 16.3: Stellar ring occultation (star is simulated in this image).

passes close behind the planet, and therefore would be incapable of producing a coherent downlink. For this reason, some spacecraft are equipped with an Ultra Stable Oscillator (USO) in a temperature-controlled "oven" on the spacecraft which is capable of providing a fairly stable downlink frequency for a short time when an uplink is not available. Calibrations, of course would, be accomplished shortly before and after the occultation.

The first occultation experiment was proposed when JPL was characterizing the precise refraction effects of Earth's atmosphere, with its known structure and composition, for the purpose of tracking spacecraft. The experimenter realized that measurements of the refraction effects induced by another planet's atmosphere could be used to "reverse-engineer" its structure and composition!

Extrasolar Planetary Occultations

Occultation may be one method of identifying extrasolar planets (also known as "exoplanets")[1] by measuring the slightly reduced brightness of a distant star as one of its planets happens to pass in front of it as viewed from Earth. A spacecraft is not necessarily required for this kind of measurement; it can be undertaken to some extent using Earth-based telescopes. More sensitive types of observations, though, are expected with the future Terrestrial Planet Finder spacecraft.[2]

Somewhat related to this subject, many Earth-based telescopes, and some proposed orbiting instruments, are equipped with an occulting disk (coronograph) that can block out light from the central star. This facilitates examining the star's "planetary" disk of material orbiting it. Scattered light in the instrument and/or in the Earth's atmosphere limits its effectiveness, but studies exist for orbiting a separate large occulting disk to improve experiments' effectiveness from Earth-based or orbiting telescopes.

Gravity Field Surveying

Planets are not perfectly spherical. Terrestrial planets are rough-surfaced, and most planets are at least slightly oblate. Thus they have variations in their mass concentrations, sometimes associated with mountain ranges or other features visible on the surface. A gravity field survey, as introduced in Chapter 8, identifies local areas of a planet that exhibit slightly more or slightly less gravitational attraction. These differences are due to the variation of mass distribution on and beneath the surface.

There are two reasons for surveying the gravity field of a planet. First, highly accurate navigation in orbit at a planet requires a good model of variations in the gravity field, which can be obtained by such a survey. Second, gravity field measurements have the unique advantage of offering scientists a "view" of mass distribution both at and below the surface. They are extremely valuable in determining the nature and origin of features identifiable in imaging data. JPL has pioneered the field of mapping planetary mass concentrations. Application of these techniques to Earth helps geologists locate petroleum and mineral deposits, as well as provide insight to geological processes at work.

To obtain gravity field data, a spacecraft is required to provide only a downlink carrier signal coherent with a highly stable uplink from the DSN. It may be modulated with telemetry, command and ranging data, or it may be unmodulated.

After the removal of known Doppler shifts induced by planetary motions and the spacecraft's primary orbit, and other factors, the residual Doppler shifts are indicative of miniscule spacecraft accelerations resulting from variations in mass distribution at, and below, the surface of the planet. The gravity feature size that can be resolved is roughly equal to the spacecraft's altitude; with a 250-km altitude, a spacecraft should resolve gravity features down to roughly 250 km in diameter.

With an X-band (3.6 cm wavelength) uplink received at a spacecraft, and a coherent X-band downlink, spacecraft accelerations can be measured to tens of micrometers per second squared. This translates to a sensitivity of milligals in a planetary gravity field. (One gal represents a gravitational acceleration of 1 cm/s^2).

The most accurate and complete gravity field coverage is obtained from low circular orbit. Mars Global Surveyor is conducting a gravity field survey from circular orbit as one of its first-priority investigations. Magellan's orbit was elliptical during its primary mission, and meaningful gravity data could

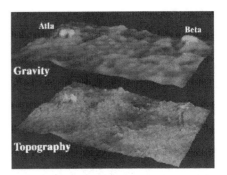

Figure 16.4: Venus gravity field vs. topography. Magellan's 4th Venus mapping cycle was dedicated to collecting gravity data. This computer-generated perspective compares gravity and topography over a region 12,700 by 8,450 km from longitudes 180° to 300° east and latitudes 40° north to 40° south. The highlands of Beta Regio and Atla Regio, sites of rifting and large volcanoes, have corresponding high topography and high gravity. These are interpreted to be sites where hot mantle material is upwelling, forming "hot spots," similar to areas on Earth such as Hawaii. The two gravity highs at Atla correspond to the volcanoes Maat Mons and Ozza Mons. Gravity anomalies at Atla and Beta are the largest on Venus and may be sites of relatively young geologic features. In contrast to Earth, there is a near-perfect correlation between gravity and topography with anomalies, both positive (light) and negative (dark), correlated with topographic highs and lows. This is interpreted to indicate that, relative to Earth, formation of features on Venus is more strongly linked to fluid motions in the mantle.

be taken only for the portion of the orbit that was plus and minus about 30 degrees true anomaly from periapsis, which occurred at about 10 degrees north latitude. After aerobraking (see below) to a low circular orbit, Magellan conducted a high-resolution gravity field survey of the entire planet.

Atmospheric Entry and Aerobraking

Aerobraking is the process of decelerating by converting velocity mostly into heat via supersonic compression in a planetary atmosphere. Galileo's atmospheric probe (see page 137) is a typical example of an atmospheric entry and aerobraking mission. The probe was designed with an aeroshell that sustained thousands of degrees of heat as it entered the Jovian atmosphere. In fact its aeroshell reached a higher temperature than the Sun's photosphere. It decelerated at hundreds of Gs, until it reached a

Figure 16.5: Galileo's Jupiter Atmospheric Probe.

speed where its parachute became effective. At that time, the spent aeroshell was discarded, and the probe successfully carried out its experiments characterizing Jupiter's upper atmosphere.

The Magellan spacecraft was not designed for atmospheric entry. However, the periapsis altitude of Magellan's orbit was lowered, by the use of propulsive maneuvers, into the upper reaches of Venus's atmosphere near 140 km above the surface. This is still high above the cloudtops, which are at about 70 km. Flying at this altitude induced deceleration via atmospheric friction during the portion of the spacecraft's orbit near periapsis, thus reducing the height to which it could climb to apoapsis (recall the discussions in Chapter 3). The solar array, consisting of two large square panels, was

Figure 16.6: Magellan's Atmospheric Plunge.

kept flat-on to the velocity vector during each pass through the atmosphere, while the HGA trailed in the wind. The solar array reached a maximum of 160° C, and the HGA a maximum of 180° C. After approximately 70 Earth days and one thousand orbits of encountering the free molecular flow, and decelerating a total of about 1250 m/sec, the apoapsis altitude was lowered to a desirable altitude. The periapsis altitude was then raised to achieve a nearly circular orbit. The objectives of this aerobraking experiment were to demonstrate the use of aerobraking for use on future missions, to characterize the upper atmosphere of Venus, and to be in position to conduct a full-planet gravity field survey from a nearly circular orbit.

Descent and Landing

Landing on a planet is generally accomplished first by aerobraking while entering the planet's atmosphere under the protection of an aeroshell. From there, the lander might be designed to parachute to the surface, or to use a propulsion system to soft-land, or both, as did the Viking landers on Mars. In addition to aeroshell, parachutes and propulsion, the Mars Pathfinder spacecraft used airbags to cushion its impact (see page 138). This technique was repeated with the Mars Exploration Rovers, Spirit and Opportunity.[3] The Soviet Venera spacecraft parachuted to the surface of Venus by means of a small rigid disk integral with the spacecrafts' structure which helped slow their descent sufficiently through the very dense atmosphere. A crushable foot pad absorbed the energy from their final impact on the surface.

Even though Huygens enjoyed the immense fortune of surviving impact with the surface of Titan, and returning data for over 90 minutes, the mission was classified as an atmospheric probe, not a lander.[4] Its prime mission was to collect data while descending for more than two hours through the atmosphere.

JPL's Surveyor missions landed on the Moon via propulsive descent, coming to rest on crushable foot pads at the lunar surface.[5]

Figure 16.7: JPL's Surveyor 3 on the Moon.

For a possible future mission, some members of the international science community desire to land a network of seismometer-equipped spacecraft on the surface of Venus to measure seismic activity over a period of months or years.

Balloon Tracking

Once deployed within a planet's atmosphere, having undergone atmospheric entry operations as discussed above, a balloon may ride with the wind and depend on the DSN to directly track its progress, or it can use an orbiting spacecraft to relay its data to Earth. In 1986, DSN tracked the Venus balloons deployed by the Soviet Vega spacecraft when it was on its way to encounter comet Halley. The process of tracking the balloon across the disc of Venus yielded data on the circulation of the planet's atmosphere.

The planned Mars Balloon is designed to descend to just above the surface. Carrying an instrument package, including a camera, within a long, snake- or rope-like structure, it will rise and float when heated by the daytime sunlight, and will sink and allow the "rope" to rest on the surface at night. In this way it is hoped that the balloon package will visit many different locations pseudo-randomly as the winds carry it. In doing so, it will also yield information on atmospheric circulation patterns. The Mars balloon is designed jointly by Russia, CNES, and The Planetary Society, a public non-profit space-interest group in Pasadena. It will depend upon an orbiting spacecraft to relay its data home.

The Mars Global Surveyor spacecraft carries radio relay equipment (see Chapter 11) designed to relay information from landers, surface penetrators, balloons, or Mars aircraft. The Mars Odyssey and Mars Express orbiters are similarly equipped. Future Mars orbiting spacecraft will also have relay capability, as did two Mars-bound spacecraft that were lost: Mars Observer in 1993 and Mars Climate Orbiter in 1999.

Sampling

One of the major advantages of having a spacecraft land on the surface of a planet is that it can take direct measurements of the soil. The several Soviet Venera landers accomplished this on the 900 F surface of Venus, and the Viking landers accomplished this on the surface of Mars. Samples are taken from the soil and transported into the spacecraft's instruments where they are analyzed for chemical composition, and the data are relayed back to Earth.

This is all good, but the scientific community really would like to directly examine samples in the laboratory. A robotic sample return mission to Mars could be in the future.[6] NASA has conducted long-range studies and technology development and may continue to do so for planning purposes. Several different scenarios are envisioned for accomplishing this, some of which would include a rover to go around and gather up rock and soil samples to deposit inside containers aboard the

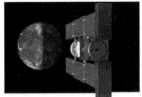

Figure 16.8: Stardust returning comet samples back to Earth.

return vehicle. Interest in returning samples from Mars has heightened recently in the wake of discoveries including recently flowing springs, and ancient lake beds on Mars. Such samples would be examined for fossil evidence of life forms.

Sampling of cosmic dust in the vicinity of the Earth has also become an endeavor of great interest, since interplanetary dust particles can reveal some aspects of the history of solar system formation. Space shuttle experiments have so far been successful at capturing three 10 m particles from Earth orbit, one intact.

Launched in 1999, Stardust collected samples of material from the coma of Comet Wild 2 in January 2004, for return to Earth January 15, 2006.[7] The return capsule's entry, descent, and landing on Earth were flawless, and scientific analysis of its thousands of samples is underway, with help from members of the public being sought. Genesis succeeded in returning samples of the solar wind, despite its unexpectedly hard landing on Earth in September 2004, when its parachute failed to deploy.[8]

A spacecraft isn't always needed if you want to collect interplanetary material. Dust from interplanetary space rains continuously into Earth's atmosphere, which slows it gently because of the particles' low mass. In 1998 the NASA Dryden Flight Research Center flew one of its ER-2 high-altitude research aircraft with an experiment for the Johnson Space Center that collected high-altitude particulate matter – or "cosmic dust" – on two collector instruments mounted on pods under the wings.[9]

Notes

[1] http://www.public.asu.edu/~sciref/exoplnt.htm
[2] http://planetquest.jpl.nasa.gov/TPF

[3]http://www.jpl.nasa.gov/missions/mer
[4]http://sci.esa.int/huygens
[5]http://nssdc.gsfc.nasa.gov/planetary/lunar/surveyor.html
[6]http://www.esa.int/export/esaCP/SEML90XLDMD_index_0.html
[7]http://stardust.jpl.nasa.gov
[8]http://genesismission.jpl.nasa.gov
[9]http://www.nasa.gov/centers/dryden/research/ER-2/

Chapter 17

Extended Operations Phase

> **Objectives:** Upon completion of this chapter, you will be able to cite examples of completion of a mission's primary objectives and obtaining additional science data after their completion. You will consider how depletion of resources contributes to the end of a mission, identify resources that affect mission life, and describe logistics of closeout of a mission.

Completion of Primary Objectives

A mission's primary objectives are spelled out well in advance of the spacecraft's launch. The efforts of all of the flight team members are concentrated during the life of the mission toward achieving those objectives. A measure of a mission's success is whether it has gathered enough data to complete or exceed its originally stated objectives.

During the course of a mission, there may be inadvertent losses of data. In the case of an orbiter mission, it might be possible to recover the losses by repeating observations of areas where the loss was sustained. Such data recovery might require additional time be added to the portion of a mission during which its primary objectives are being achieved. However, major data losses and their recovery are usually planned for during mission design. One predictable data loss occurs during superior conjunction, when the Sun interferes with spacecraft communications for a number of days. In Figure 17.1, missing Magellan radar data appears as swaths from pole to pole (arrows) which have been

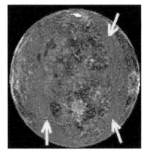

Figure 17.1: Gaps in Magellan's Venus data were filled in during mission extensions.

filled in with lower-resolution data from the Pioneer 12 mission. Magellan later recovered the missing high-resolution data.

Additional Science Data

Once a spacecraft has completed its primary objectives, it may still be in a healthy and operable state. Since it has already undergone all the efforts involved in conception, design and construction, launch, cruise and perhaps orbit insertion, it can be very economical to redirect an existing spacecraft toward accomplishing new objectives and to retrieve data over and above the initially planned objectives. This has been the case with several JPL spacecraft. It is common for a flight project to have goals in mind for extended missions to take advantage of a still-viable spacecraft in a unique location when the original funding expires.

Voyager (see pages 78 and 135) was originally approved as a mission only to Jupiter and Saturn. But Voyager 2's original trajectory was selected with the hope that the spacecraft might be healthy after a successful Saturn flyby, and that it could take advantage of that good fortune. After Voyager 1 was successful in achieving its objective of reconnaissance of the Saturnian system, including a tricky solar occultation of Titan and associated observations, Voyager 2 was not required to be used solely as a backup spacecraft to duplicate these experiments. Voyager 2's trajectory to Uranus and Neptune was therefore preserved and successfully executed. Approval of additional funding enabled making some necessary modifications, both in the ground data system and in the spacecraft's onboard flight software, to continue on to encounter and observe the Uranus and Neptune systems.

By the time Voyager 2 reached Uranus after a five-year cruise from Saturn, it had many new capabilities, such as increased three-axis stability, extended imaging exposure modes, image motion compensation, data compression, and new error-correction coding.

In 1993, after 15 years of flight, Voyagers 1 and 2 both observed the first direct evidence of the long-sought-after heliopause. They identified a low frequency signature of solar flare material interacting with the heliopause at an estimated distance of 40 to 70 AU ahead of Voyager 1's location, which was 52 AU from the Sun at the time.

After fulfilling its goal of mapping at least 70% of the surface of Venus, the Magellan mission went on with more than one mission extensions, eventually to accomplish special stereo imaging tests, and interferometric observation tests. Mapping coverage reached over 99% of the surface.

Rather than abandon the spacecraft in orbit, the Magellan Project applied funding which had been saved up over the course of the primary mission to begin an adventurous transition experiment, pioneering the use of aerobraking to attain a nearly circular Venusian orbit, and a low-latitude gravity survey was completed. All of these accomplishments far exceeded the mission's original objectives.

Orbiting Relay Operations

As mentioned in the last chapter, some Mars orbiting spacecraft are equipped with radio relay capability intended to receive uplink from surface or airborne craft. Typically such relay equipment operates at UHF frequencies (see page 93).

In order to serve as a relay, at least some of the orbiter's own science data gathering activities have to be reduced or interrupted while its data handling and storage subsystems process the relay data.[1] This may or may not present an undesirable impact to the mission's ability to meet its primary objectives. Relay service, then, is a good candidate for extended mission operations.[2] Since relay service entails neither keeping optical instruments pointed, nor flying a precise ground track, the demand on the attitude control and propulsion systems is minimal, and a little propellant can go a long way. The demand on other subsystems, such as electrical supply, can also be reduced in the absence of other science instruments to operate.

End of Mission

Resources give out eventually. Due to the age of their RTGs in 2000, the Pioneer 10 and 11 spacecraft,[3] plying the solar system's outer reaches, faced the need to turn off electrical heaters for the propellant lines in order to conserve electrical power for continued operation of science instruments. Doing so allowed the propellant to freeze, making it impossible to re-thaw for use in additional spacecraft maneuvers. The spacecraft were still downlinking science data while Earth eventually drifted away from their view, and over the following months contact was lost forever.

Voyagers 1 and 2 (see page 135) continue to make extraordinary use of their extended mission. They are expected to survive until the sunlight they observe is too weak to register on their sun sensors, causing a loss of attitude reference. This is forecast to happen near the year 2015, which may or may not be after they have crossed the heliopause. Electrical energy from their RTGs may fall below a useable level about the same time or shortly thereafter. The spacecraft's supply of hydrazine may become depleted sometime after that, making continued three-axis stabilization impossible.

Pioneer 12 ran out of hydrazine propellant in 1993, and was unable to further resist the slow decay of its orbit about Venus, resulting from friction with the tenuous upper atmosphere.[4] It entered the atmosphere and burned up like a meteor after fourteen years of service.

Components wear out and fail. The Hubble Space Telescope has been fitted with many new components, including new attitude-reference gyroscopes, to replace failed and failing units. Two of Magellan's attitude-reference gyroscopes had failed prior to the start of the transition experiment, but of course no replacement was possible. To date, a JPL mission has not been turned off because of lack of funding.

Once a mission has ended, the flight team personnel are disbanded, and the ground hardware is returned to the loan pool or sent into long-term storage. Sometimes it is possible to donate excess computers to

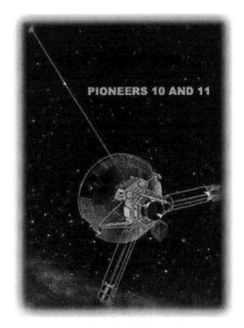

Figure 17.2: Pioneer 10 and 11 were commanded to shut down in stages as electrical power declined. Image courtesy NASA Ames.

schools. Oversubscribed DSN resources are freed of contention from the terminated mission, and the additional tracking time allocations can be made available to missions currently in their prime.

While layoffs are not uncommon, many personnel from a disbanded flight team are assigned by their JPL section management to new flight projects to take advantage of valuable experience gained. Interim work is often available within the Section itself. Many Viking team members joined the Voyager mission after Viking achieved its success at Mars in the late 1970s. Many of the Voyager flight team members joined the Magellan project after Voyager's last planetary encounter ended in October 1989.

Other ex-Voyager people joined the Galileo and Topex/ Poseidon missions. Some ex-Magellan people have worked on Cassini, Mars Global Surveyor, Mars Pathfinder, Spitzer, and Mars Exploration Rover. Mission's end also provides a convenient time for some employees to begin their retirement, and for new employees to be hired and begin building careers in interplanetary exploration.

Notes

[1]http://science.ksc.nasa.gov/mirrors/jpl/pathfinder/msp98/msss/mars_relay
[2]http://www.esa.int/export/SPECIALS/Mars_Express/SEM5S9W4QWD_0.html
[3]http://www.nasa.gov/mission_pages/pioneer
[4]*ibid.*

Chapter 18

Deep Space Network

> **Objectives:** Upon completion of this chapter, you will be able to describe the Deep Space Network's seven data types and trace data flow. You will be able to describe the three Deep Space Communications Complexes, and compare five types of Deep Space Stations. You will be able to describe advantages of arraying and cite planned improvements in the DSN.

The **NASA Deep Space Network, DSN,** is an international network of facilities managed and operated by JPL. The DSN supports interplanetary spacecraft missions, radio astronomy, radar astronomy, and related observations for the exploration of the solar system and the universe. The DSN also supports selected Earth-orbiting missions.

The DSN is the largest and most sensitive scientific telecommunications system in the world. It consists of three deep-space communications complexes, DSCCs, placed approximately 120 degrees apart around the world: at Goldstone near Barstow in California's Mojave Desert; at Robledo near Madrid, Spain; and at Tidbinbilla near Canberra, Australia. The Network Operations Control Center, NOCC, is in building 230 at JPL. The strategic placement of the DSCCs permits constant observation of spacecraft on interplanetary missions as the Earth rotates.

Figure 18.1: One of DSN's 34 m aperture Beam Waveguide Deep Space Stations.

The DSN enjoys a rich history[1] that is closely intertwined with the history of the space age and the cutting edge of development in telecommunications technology.

253

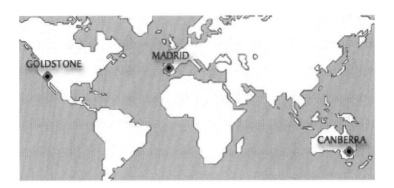

The Seven DSN Data Types

The DSN is an extremely complex facility, but it becomes more easily comprehensible if you recognize its seven data types, as a context for learning about DSN subsystems, and how they relate to each other. In the past, each of these seven data types was associated with a separate DSN system. Today, thanks to the Network Simplification Program, these have been consolidated into two major DSN subsystems: The Uplink Tracking and Command Subsystem, UPL, and The Downlink Tracking & Telemetry Subsystem, DTT.

Here is a brief discussion of the seven DSN data types that are processed in the UPL and DTT:

1. Frequency & Timing Data Type, F&T

Any computer system, whether desktop or supercomputer, has an internal clock that directs every step of the computer's operations. F&T is the DSN's "internal clock." With precision and accuracy that are at the forefront of world class frequency and timing science, the Frequency & Timing Subsystem is essential to nearly every part of the DSN, enabling the other six data types to exist.

At the heart of F&T are four frequency standards of which one is prime and the other three are backups. These include the hydrogen masers mentioned in Chapter 10 (see page 146), and cesium frequency standards. The master clock assembly produces time codes using the frequency standard as a reference. Every subsystem across the DSN, and nearly every assembly have an input of F&T data in the form of a reference frequency and/or time codes. Those subsystems having time code inputs interface via time code translators, TCTs.

F&T synchronization is managed among all three DSCCs and JPL by keeping track of offsets in fractions of microseconds resulting from comparison of local F&T data with reference pulses received from Global Positioning System, GPS, satellites.

2. Tracking Data Type, TRK

The TRK data type includes Doppler, ranging, predicts, and DSN antenna control.

Measurement of the Doppler shift on a spacecraft's coherent downlink carrier allows determination of the line-of-sight component of the spacecraft's velocity. Routine measurement precision is on the order of fractions of a millimeter per second.

Ranging tones uplinked and transponded by a spacecraft enable navigators to determine an average distance to and from the spacecraft, with a routine precision of about one meter.

Navigators use Doppler and range measurements to determine a spacecraft's trajectory, and to infer gravity fields of bodies that affect the spacecraft. Navigation team members create ephemeris files that the DSN uses to generate antenna pointing predicts and frequency predicts for uplink and downlink. Predicts are sent to DSN sites to enable acquisition and following of the spacecraft.

3. Telemetry Data Type, TLM

The word Telemetry is derived from the Greek "tele" (far off), and "metron" (measure). A spacecraft produces digital data to represent engineering measurements, such as the temperatures of parts of the spacecraft, and science data, such as images from its cameras.

The spacecraft places symbols on its radio frequency downlink to represent the ones and zeroes that make up this data. The DSN Downlink Tracking & Telemetry subsystem recreates the spacecraft's digital data bit for bit by recognizing the downlinked symbols and decoding them. DSN then delivers the TLM data to the flight project for display, distribution, storage, and analysis, supporting spacecraft engineering management and eventual publication of scientific results.

4. Command Data Type, CMD

Flight projects send digital data to the spacecraft via the DSN Uplink Tracking & Command subsystem. Like telemetry-in-reverse, digital bits generated by the flight project are sent as CMD data to the spacecraft, which is able to recognize the bits as either flight software to load into its on-board computers, or as commands to control the spacecraft's activities.

5. Monitor Data Type, MON

MON data reports on the operation and performance of the DSN itself. The DSN Network Monitor & Control subsystem (NMC) collects data from assemblies throughout its subsystems. This MON data is used in various locations: within the DSCC to watch and control its own activities; at the Network Operations and Control Center at JPL for managing and advising DSN operations, and in flight projects to help with realtime coördination of operations.

Flight projects typically select a subset of MON data to distribute and store along with TLM data to provide indications of, for example, the strength of the spacecraft's signal as received by DSN at any given time.

6. Radio Science Data Type, RS

As mentioned in Chapter 8 (see page 122), RS experiments use the spacecraft radio and the DSN together as a science instrument. RS investigators remotely control equipment in the DSN such as the Radio Science Receivers, RSR, to capture and record data on the attenuation, scintillation, refraction, rotation, Doppler shifts, and other direct modifications of a spacecraft's radio signal as it is affected by the atmosphere of a planet, the sun, moons, or by structures such as planetary rings or gravitational fields.

Unlike the closed-loop receivers used by TRK and TLM, RS uses open-loop receivers and spectrum processing equipment. Rather than lock onto one discrete frequency, the open-loop equipment can observe a range of frequencies.

The JPL Radio Science System Group has an informative website.[2]

7. Very Long Baseline Interferometry Data Type, VLBI

VLBI can be applied to a number of investigations. Two or more widely separated DSN stations observe the same spacecraft, or a quasar, at the same time, using open-loop receivers, and record their data. The recorded data is taken to a special-purpose computer called a correlator for processing to produce a cross-correlation fringe pattern. Further analysis can precisely determine the relative position of the antennas. This investigation is called geodesy. With the antenna positions known precisely, VLBI can precisely determine the position of a spacecraft. VLBI can also produce synthetic aperture results such as images of astronomical objects (see page 198. There is a technical tutorial on VLBI online.[3]

The DSN Facilities

In addition to the three DSCCs and the NOCC, the DSN also includes the following:

- The Demonstration Test Facility at JPL known as DTF-21 where spacecraft-to-DSN compatibility is demonstrated and tested prior to launch,

- The Merrit Island facility MIL-71 in Florida, which supports launches, and

- The Ground Communications Facility, GCF, which connects them all with voice and data communications. The GCF uses land lines, submarine cable, terrestrial microwave, and communications satellites.[4]

A Closer Look at the DSCCs

All three DSCCs have generally the same makeup, although Goldstone, GDSCC, being closest to JPL, has some additional antennas, as well as research and development facilities not found at the others: Madrid, MD-SCC, or Canberra, CDSCC. Each DSCC has the following:

- A number of Deep Space Stations, DSSs. Each DSS comprises a high-gain, parabolic-reflector steerable antenna dish, and its associated front-end equipment such as low-noise amplifiers and transmitters. That's a DSS pictured near the top of this page. The DSSs are divided, according to their aperture size, into groups called subnets, for example the 70m subnet, the 34m subnet, and the 26m subnet. (Subnets are further identified by antenna type, as described on the next page.)

- The signal processing center, SPC. The SPC connects with all the DSSs at the DSCC, and houses the operations personnel along with the computers and other equipment within the UPL and the DTT that handle the seven DSN data types.

- Administrative offices, a cafeteria, and a visitor center.

The remainder of this chapter discusses the DSSs and their different capabilities, and describes the flow of TLM, MON, TRK, RS, and CMD data.

DSS Designations

- All the DSSs at the GDSCC in California are designated with numbers in the teens and twenties: DSS13, DSS14, DSS15, DSS24, DSS25, DSS26, etc.

- All the DSSs at the CDSCC in Australia are designated with numbers in the thirties and forties: DSS34, DSS43, DSS45 and DSS46.

- All the DSSs at the MDSCC in Spain are designated with numbers in the fifties and sixties: DSS54, DSS55, DSS65, and DSS63.

The 70 m Subnet:
DSS14, DSS43, DSS63

The three 70 m DSSs were originally built as 64 m diameter antennas. The first, GDSCC's DSS-14, also known as the Mars Station for its support of Mariner-4, began operation in 1966. All three were expanded to 70 m diameter from 1982 to 1988 to increase their sensitivity to support Voyager 2's encounter with Neptune.

The 70 m DSSs are used for deep-space mission support, radio astronomy, and VLBI. The 70 m DSS14 at GDSCC is also used for radar astronomy, which is known as Goldstone Solar System Radar, GSSR.

Figure 18.2: Deep Space Station 43 in Canberra, Australia.

The 70 m subnet DSSs support both X-band and S-band uplink and downlink.

The 34 m HEF Subnet: DSS15, DSS45, DSS65

HEF stands for High-Efficiency. This subnet was installed to replace the older 34 m Standard, STD, subnet. The 34 m STD subnet DSSs had a polar-axis, or HA-DEC, design (see page 53) and were originally built with 26 m diameter reflectors and later upgraded to 34 m. The upgrade required repositioning the entire DSS up on concrete footings so that the reflector could point to low elevations without striking the ground. The 34 m STD DSSs at CDSCC (DSS42) and at MDSCC (DSS61) have been dismantled.

Figure 18.3: Deep Space Station 15 at Goldstone, California.

The GDSCC 34 m STD (DSS12) was converted to an educational resource and renamed the Goldstone-Apple Valley Radio Telescope, GAVRT.

The 34 m HEF subnet was designed with the more efficient azimuth-elevation mounting and a 34 m reflector from the start, with a precision-shaped surface for maximum signal-gathering capability at X-band radio frequencies. The 34 m HEF subnet supports mostly deep space missions, but may occasionally support a mission in high Earth orbit.

The 34 m HEF subnet DSSs support X-band uplink and downlink, and S-band downlink.

The 34 m BWG Subnet: DSS24, DSS25, DSS26, DSS34, DSS54

BWG stands for Beam Wave Guide. These, the newest of the DSSs to be designed and implemented, can be recognized by the hole in the middle of their main reflectors where on other DSSs you'd find a feed cone full of microwave equipment. The BWG design does not require sensitive equipment to be mounted in a cramped, moveable feedhorn or room mounted beneath the main reflector, where it is difficult to access for maintenance, repair, and upgrade. Instead, the BWG DSS directs the microwave beam through waveguides via five precision RF mirrors down into a basement room where equipment can be stably mounted on the floor and on balconies, with plenty of room for access.

Figure 18.4 shows two of the beam waveguide reflectors inside the antenna mounting. These two reflectors maintain the microwave beam as the antenna moves around in azimuth. Similar reflectors, located higher up in the structure, permit freedom of movement in elevation. In Figure 18.1 on page 253, note the hole in the center of the main reflector where, on other DSS designs, there is a feed-cone containing electronics.

The 34 m BWG subnet supports mostly deep space missions, but may occasionally support a mission in high Earth orbit.

Figure 18.4: Microwave Mirrors in the basement of a 34 m Beam Waveguide DSS.

The 34 m BWG subnet DSSs generally support both X-band and S-band (see page 93) uplink and downlink, although this is not true for all of the BWG DSSs at GDSCC. Some at GDSCC also have Ka-band uplink and downlink capability.

There is a fascinating time-lapse movie available online that illustrates the construction process of DSS55, which is a new 34 m BWG DSS in Spain.[5]

The 26 m Subnet:
DSS16, DSS46, DSS66

The 26 m diameter subnet is no longer considered part of the DSN. These antennas were used for rapidly tracking Earth-orbiting spacecraft. They were originally built to support the Apollo lunar missions between 1967 and 1972.

The XY-mounted 26 m DSSs (see page 54) were originally part of the Spaceflight Tracking and Data Network, STDN, operated by NASA's Goddard Space Flight Center in Greenbelt, Maryland. They were integrated into the DSN in 1985 when it needed them to track spacecraft in highly elliptical Earth orbits.

Arraying

A powerful technique for obtaining higher sensitivity in support of distant spacecraft is arraying. Arraying means electronically combining the signals coming in from two or more DSSs, either at the same DSCC, at two different DSCCs, or with a non-DSN radio telescope. This increases the effective aperture, strengthens reception of the spacecraft's weak signal, and permits the spacecraft to be able to downlink data at a higher rate.[6]

Arraying was conducted between DSS43 and the Parkes radio telescope in Australia in support of the 1986 Voyager 2 encounter with Uranus. The 27-antenna Very Large Array in Socorro New Mexico was arrayed with DSS14 in support of Voyager's 1989 encounter with Neptune. Intercontinental arraying was accomplished in support of the Galileo mission at Jupiter.

Arraying four 34 m DSSs can provide the equivalent of one 70 m DSS. As many as eight antennas can be arrayed at a DCSS.

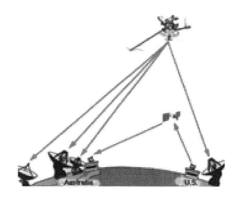

Figure 18.5: Intercontinental DSN Array.

Advances in the DSN

New antennas are added to the DSN, as budget and circumstance permit, to help support the growing demands for communication and navigation, and to help act as a backup in case of any need for extended periods of maintenance or repair for existing DSSs. Recently added antennas are of the 34 m BWG type.

Future plans call for continuing to add 34 m BWG antennas to each DSCC. This was decided after carefully studying the concept of using a large number of 12 m aperture units, that could be arrayed to achieve performance equivalent to the giant 34 m and 70 m DSSs. One paper describing the concept is by Mark S. Gatti, and is titled, *A Phased Array Antenna for Deep Space Communications*[7]

There have been studies of the concept of having a communications system located in orbit at Mars. Rather than have all spacecraft communicate individually with antennas of the DSN, such a network would help consolidate the communications and navigation tasks at the red planet. Unfortunately, the first real component of such a network, the Mars Telecommunications Orbiter, was cancelled in July of 2005 due to budget constraints. The Mars Reconnaissance Orbiter (MRO) spacecraft will have to serve as the main relay for data to be returned by the Mars Science Laboratory

(MSL), a highly capable surface rover scheduled for launch in 2009.[8]

Beginning here, we'll follow a typical communications link from a spacecraft into the DSS antenna, through the DSN Downlink Tracking & Telemetry subsystem (DTT), and on to JPL. Also, command data is traced from JPL to the DSCC and out the DSS antenna toward the spacecraft via the DSN Uplink Tracking & Command subsystem. The first diagram, Figure 18.6, shows equipment located within a DSS. The second diagram, Figure 18.7, shows equipment located within the SPC.

Data Flow at the DSCC

Downlink radio-frequency energy (RF) enters the DSS antenna reflector shown in black in Figure 18.6, which represents a 34 m Beam Waveguide (BWG) DSS, and proceeds down the center line, which represents waveguides. Along this initial path is where the five reflectors of a BWG would be located, directing the RF into the basement where the rest of the equipment is located. With other DSSs, all the equipment in Figure 18.6 is located in the feed-horn and just below the reflector, where it all moves with the reflector as it tracks the spacecraft.

The arrow on the left of Figure 18.6 indicates antenna control signals going to the DSS antenna equipment from the DTT in the SPC. All the other components in this diagram belong to the Microwave subsystem (UWV).

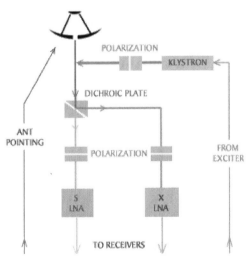

The center line comes to a dichroic plate, also called a dichroic mirror. RF at one frequency, for example S-band, passes through the plate, to the gold colored path below, while RF of another frequency, for example X-band, reflects off to follow

Figure 18.6: The BWG microwave system.

waveguides in another path to the right. Some DSSs can also select Ka-band or other bands of RF. The desired polarization is selected using filters. This might be right-hand circular polarization, RCP, or left-hand, LCP, or none.

The RF of each band, S and X in this example, goes to a low-noise amplifier, LNA. The LNA used depends on what is installed at a particular DSS and the needs of the user. It may be a cryogenically cooled amplifier called a maser (an acronym from "Microwave Amplification by Stimulated Emission of Radiation"), or it may be a solid-state device called a high-electron-mobility transistor, HEMT. The function of the LNA is to amplify a band of RF while introducing an absolute minimum amount of noise. DSN's masers are cooled with liquid helium to keep RF noise down.

An amplified band of RF leaves the LNA and is directed toward the receivers, which are shown in the next diagram. Before leaving the DSS, the downlink RF is converted to a lower frequency signal known as the Intermediate Frequency (IF) signal. The IF signal can be carried more conveniently, by coaxial cable or fiber optics for example, than the RF signal which typically needs waveguides.

Before getting to the next diagram, notice the red line in Figure 18.6 labelled "From Exciter" coming up to the klystron on the right. This represents the uplink signal to be amplified by the klystron, which is a microwave power amplifier vacuum tube. The signal is generated by the exciter (part of the receiver) based on a reference frequency provided by FTS, and other inputs to be discussed later.

Polarization of the klystron's output is selected by a filter to match the spacecraft's receiver. The klystron's output illuminates the DSS's antenna reflector so it can be seen by the spacecraft. Most klystrons have to be actively cooled by refrigerated water or other means. The klystron, its high-voltage-DC, high-current power supply, and its cooling apparatus are collectively known as the Transmitter subsystem, TXR, or XMTR.

Looking at the lower diagram, Figure 18.7 which represents equipment in the SPC, the IF signal from the LNA in the DSS enters at the top. If two or more LNAs are operating, the path would be multiple. The diagram shows LNAs for X-band and S-band, but there are also Ka-band LNAs installed. Depending on operations, the LNA may feed either a closed-loop receiver,an open-loop receiver, or both.The switches in the diagram show there's an operational choice. An open-loop receiver is used for radio science, and also for VLBI.

The open-loop receivers select a band of frequencies to amplify for further processing and storage by RS or VLBI equipment. VLBI equipment typically outputs data to tapes that are delivered to a correlator at a different location. Radio science equipment, controlled remotely from JPL, will typically output its high-volume data online for transmission to JPL via the GCF, indicated by the block at the bottom of the diagram, or by other carriers.

RF from the LNA can also go to the closed-loop receiver. DSN uses its highly advanced receiver known as the Block-V receiver (V is the Roman numeral five), BVR. The BVR is part of the Receiver & Ranging Processor (RRP) in one of several Downlink Channels (DC) in the DTT subsystem. In the BVR a single frequency, the spacecraft's downlink, is selected and amplified. If there are any subcarriers carrying telemetry or ranging data, they are detected here, and if symbols are present on the

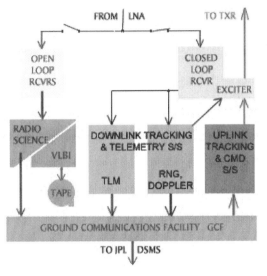

Figure 18.7: The SPC signal and data paths.

carrier or subcarriers, they are recovered within the BVR for decoding and further processing.

At the discretion of the Ace, a program known as conscan may be invoked. Conscan, which stands for conical scanning, observes the closed-loop receiver's signal strength and adjusts the antenna pointing via the DTT. The antenna constantly moves in small, tightening circles as it optimizes its pointing. Conscan must be disabled when the spacecraft's signal changes or disappears. It is not desirable to conscan during VLBI or RS operations due to the variations it induces in signal level. RS does, however, have a feature called monopulse to optimize Ka-band reception. Monopulse creates records of the adjustments it induces that can be accounted for in data analysis.

In the RRP, the downlink's Doppler shift is measured and compared with the predicted Doppler shift. The difference is called the Doppler residuals. If there are ranging symbols on the downlink, they are processed within this subsystem as well. The ranging and Doppler data is passed to the navigators at JPL via the GCF.

If there are telemetry symbols present on the downlink, they are further processed within the DC. First, if applicable to the particular spacecraft, they are Viterbi-decoded to recover data bits from the convolutionally coded symbol stream.[9] The assembly that does this is the maximum-likelihood convolutional decoder, MCD. If any other coding, such as Reed-Solomon is present, it can be decoded here or at JPL. The bits are then grouped into the same packages, called transfer frames, that the spacecraft had grouped them prior to downlink. Most newer spacecraft comply with the CCSDS standards for grouping bits into packets and transfer frames.[10] The TLM data is then sent to JPL via the GCF.

Command data intended to be sent to the spacecraft comes from JPL via the GCF as indicated on the right side of the diagram. The Uplink Tracking & Command subsystem (UPL) processes the data and sends the bits, on a subcarrier if applicable, to the exciter. Also, and not shown in the diagram, is a response from the UPL to JPL. The response, also designated CMD data, includes information identifying the CMD bits that have left the antenna, and reports on UPL states and operations. If operations call for placing ranging symbols and/or a ranging subcarrier on the uplink, the UPL generates and passes these signals to the exciter.

The exciter then creates a complete uplink signal with any appropriate command subcarrier, ranging subcarrier, and/or direct data modulation. It sends this signal to the transmitter, which will amplify it enough for the spacecraft to receive it across vast reaches of space.

The GCF, indicated by the lower block, uses a reliable network service, RNS, to deliver data to a central data recorder, CDR, at JPL. RNS, using the TCP/IP communications protocol, automatically replays any data that may have gotten dropped during its trip to JPL. The result, given enough time to identify and process the replays, is 100% error-free data transmission.

Some DSN subsystems are not represented in the simplified diagrams above. Frequency & Timing, for example, has inputs to all the subsystems shown. MON data is also collected from all the subsystems and sent to JPL via the GCF.

Colorful Equipment

In operations, subsystems and assemblies are called "green" when they have been operating nominally for a period of time including at least one previous tracking pass. Anything inoperable is designated "red" equipment. And anything that has been repaired and is returning to use during the present tracking pass is designated "orange."

Data at JPL

All the data streams, TLM, MON, TRK, RS, CMD, are processed and/or stored at JPL by the Deep Space Mission System, DSMS. The DSMS uses advanced software and high-performance workstations to process and route the data, to broadcast data in real time, to distribute data, to display data, to store data in online repositories for later query by users, and to archive data on permanent media.

Notes

[1] http://deepspace.jpl.nasa.gov/dsn/history/album/album.html

[2] http://radioscience.jpl.nasa.gov

[3] http://ipnpr.jpl.nasa.gov/progress_report2/42-46/46C.PDF

[4] The GCF has recently been replaced by a modern, reliable, high-speed communications system called the Data Capture and Delivery system, DCD.

[5] http://deepspace.jpl.nasa.gov/dsn/movies/DSN_Antenna_movie.mov

[6] http://deepspace.jpl.nasa.gov/dsn/antennas/arrays.html

[7] IEEEAC #1606, December 2007).

[8] http://marsprogram.jpl.nasa.gov/msl

[9] http://home.netcom.com/~chip.f/viterbi/tutorial.html

[10] http://public.ccsds.org

Part IV

REFERENCE

Units of Measure

Unit Abbreviations Frequently Found in the Text

AU – Astronomical Unit, a measure of distance, based on the mean sun-Earth distance. The International Astronomical Union defines the AU as the distance from the Sun at which a particle of negligible mass, in an unperturbed orbit, would have an orbital period of 365.2568983 days (a Gaussian year). The AU is thus defined as 1.4959787066E+11m (149,597,870.66 km).

bps – Bits per second, a measure of data rate

c – Speed of light in a vacuum, 299,792,458 m/s

G – Giga, a multiplier, $\times 10^9$ from the Latin "gigas" (giant). In the U.S., 10^9 is a billion, while in other countries using SI, 10^{12} is a billion. Giga means 10^9 everywhere. The remaining multipliers are listed in the Glossary.

g – Gram, a unit of mass (see SI units below)

Hz – Hertz, the number of cycles per second

LY – Light Year, a measure of distance, the distance light travels in one year; about 63,240 AU

m – Meter, a unit of length (USA spelling; elsewhere, metre) (see SI units below)

N – Newton, a unit of force (see SI units with special names, below)

s – Second, the SI unit of time

W – Watt, a unit of power (see SI units with special names, below)

International System of Units, SI

SI has long been the notation universally used in science and technology. It has also become the dominant language of international commerce and trade, except in the U.S.

For a comprehensive and definitive reference on all aspects of SI, as well as many other quantities and standards, please visit the National Institute

Table 1: SI Base Quantities, Unit Names, & Symbols

Base quantity	Name	Symbol
length	meter	m
mass	kilogram	kg
time	second	s
electric current	ampere	A
thermodynamic temperature	kelvin	K
amount of substance	mole	mol
luminous intensity	candela	cd

Table 2: Some Derived SI Quantities

Derived quantity	Name	Symbol
area	square meter	m^2
volume	cubic meter	m^3
speed, velocity	meter per second	m/s
acceleration	meter per second per second	m/s^2
wave number	reciprocal meter	m^{-1}
mass density	kilogram per cubic meter	kg/m^3
specific volume	cubic meter per kilogram	m^3/kg
current density	ampere per square meter	A/m^2
magnetic field strength	ampere per meter	A/m

of Standards (NIST) website (some of the information on this page has been obtained from there):

http://physics.nist.gov/cuu/Units

See also the Solar System Temperature Reference on page 19 for examples and temperature comparisons of objects and conditions in space, from absolute zero through planet temperatures, to those of stars.

Table 3: Selected SI Units with Special Names

Derived quantity	Name	Symbol	Expression
plane angle	radian	rad	-
solid angle	steradian	sr	-
frequency	hertz	Hz	-
force	newton	N	-
pressure, stress	pascal	Pa	N/m^2
energy, work, quantity of heat	joule	J	N·m
power, radiant flux	watt	W	J/s
electric charge, qty of electricity	coulomb	C	-
electric potential difference, electromotive force	volt	V	W/A
capacitance	farad	F	C/V
electric resistance	ohm	Ω	V/A
electric conductance	siemens	S	A/V
magnetic flux	weber	Wb	V·s
magnetic flux density	tesla	T	Wb/m^2
inductance	henry	H	Wb/A
Celsius temperature	degree	°C	-
luminous flux	lumen	lm	cd·sr
illuminance	lux	lx	lm/m^2

(Refer to NIST website for expressions in terms of SI base units)

A Few Handy SI-to-English Conversions

Take the number of SI units and apply the conversion to get the number of English units. For example, 2 meters equals about 6.56 feet.

Table 4: SI-to-English Conversions

Millimeters to inches:	mm	×	0.0393700787401575	=	in
Centimeters to inches:	cm	×	0.393700787401575	=	in
Meters to feet:	m	×	3.28083989501312	=	ft
Meters to yards:	m	×	1.09361329833771	=	yds
Kilometers to miles:	km	×	0.621371192237334	=	mi
Grams to ounces:	g	×	0.0352739907229404	=	oz
Kilograms to pounds:	kg	×	2.20462262184878	=	lbs
Celsius to Fahrenheit:	(°C	×	9/5) + 32	=	°F
Newtons to Pounds Force:	N	×	0.224809024733489	=	lbf

Glossary

a, A – Acceleration. a = Δ velocity / Δ time. Acceleration = Force / Mass

A – Ampere, the SI base unit of electric current.

Å – Angstrom (0.0001 micrometer, 0.1 nm).

A Ring – The outermost of the three rings of Saturn that are easily seen in a small telescope.

AAAS – American Association for the Advancement of Science.

AACS – Attitude and Articulation Control Subsystem onboard a spacecraft.

AAS – American Astronomical Society.

AC – Alternating current.

Acceleration – Change in velocity. Note that since velocity comprises both direction and magnitude (speed), a change in either direction or speed constitutes acceleration.

ALT – Altitude.

ALT – Altimetry data.

AM – Ante meridiem (Latin: before midday), morning.

am – Attometer (10^{-18} m).

AMMOS – Advanced Multimission Operations System.

Amor – A class of Earth-crossing asteroid.

AO – Announcement of Opportunity.

AOS – Acquisition Of Signal, used in DSN operations.

Aphelion – Apoapsis in solar orbit.

Apoapsis – The farthest point in an orbit from the body being orbited.

Apogee – Apoapsis in Earth orbit.

Apochron – Apoapsis in Saturn orbit.

Apojove – Apoapsis in Jupiter orbit.

Apollo – A class of Earth-crossing asteroid.

Apolune – Apoapsis in lunar orbit.

Apselene – Apoapsis in lunar orbit.

Argument – Angular distance.

Argument of periapsis – The argument (angular distance) of periapsis from the ascending node.

Ascending node – The point at which an orbit crosses a reference plane (such as a planet's equatorial plane or the ecliptic plane) going north.

ASRG – Advanced Sterling Radioisotope Thermoelectric Generator.

Asteroids – Small bodies composed of rock and metal in orbit about the sun.

Aten – A class of Earth-crossing asteroid.

Attometer – 10^{-18} meter.

AU – Astronomical Unit, based on the mean Earth-to-sun distance, 149,597,870 km. Refer to "Units of Measure" on page 269 for more complete information.

AZ – Azimuth.

△

B – Bel, a unit of ratio equal to ten decibels. Named in honor of telecommunications pioneer Alexander Graham Bell.

B Ring – The middle of the three rings of Saturn that are easily seen in a small telescope.

Barycenter – The common center of mass about which two or more bodies revolve.

Beacon – Downlink from a spacecraft that immediately indicates the state of the spacecraft as being one of several possible states by virtue of the presence and/or frequency of the subcarrier. See Chapter 10.

Bel – Unit of ratio equal to ten decibels. Named in honor of telecommunications pioneer Alexander Graham Bell.

Billion – In the U.S., 10^9. In other countries using SI, 10^{12}.

Bi-phase – A modulation scheme in which data symbols are represented by a shift from one phase to another. See Chapter 10.

BOT – Beginning Of Track, used in DSN operations.

BPS – Bits Per Second, same as Baud rate.

BSF – Basics of Space Flight (this document).

BVR – DSN Block Five (V) Receiver.

BWG – Beam waveguide 34-m DSS, the DSN's newest DSS design.

△

c – The speed of light, 299,792 km per second.

C-band – A range of microwave radio frequencies in the neighborhood of 4 to 8 GHz.

C Ring – The innermost of the three rings of Saturn that are easily seen in a small telescope.

Caltech – The California Institute of Technology.

Carrier – The main frequency of a radio signal generated by a transmitter prior to application of any modulation.

Cassegrain – Reflecting scheme in antennas and telescopes having a primary and a secondary reflecting surface to "fold" the EMF back to a focus near the primary reflector.

CCD – Charge Coupled Device, a solid-state imaging detector.

C&DH – Command and Data Handling subsystem on board a spacecraft, similar to CDS.

CCS – Computer Command subsystem on board a spacecraft, similar to CDS.

CCSDS – Consultative Committee for Space Data Systems, developer of standards for spacecraft uplink and downlink, including packets.

CDR – GCF central data recorder (obsolete).

CDS – Command and Data Subsystem onboard a spacecraft.

CDSCC – DSN's Canberra Deep Space Communications Complex in Australia.

CDU – Command Detector Unit onboard a spacecraft.

Centrifugal force – The outward-tending apparent force of a body revolving around another body.

Centimeter – 10^{-2} meter.

Centripetal acceleration – The inward acceleration of a body revolving around another body.

CGPM – General Conference of Weights and Measures, Sevres France. The abbreviation is from the French. CGPM is the source for the multiplier names (kilo, mega, giga, etc.) listed in this document.

Chandler wobble – A small motion in the Earth's rotation axis relative to the surface, discovered by American astronomer Seth Carlo Chandler in 1891. Its amplitude is about 0.7 arcseconds (about 15 meters on the surface) with a period of 433 days. It combines with another wobble with a period of one year, so the total polar motion varies with a period of about 7 years. The Chandler wobble is an example of free nutation for a spinning non-spherical object.

Channel – In telemetry, one particular measurement to which changing values may be assigned. See Chapter 10.

CIT – California Institute of Technology, Caltech.

Clarke orbit – Geostationary orbit.

CMC – Complex Monitor and Control, a subsystem at DSCCs.

CMD – DSN Command System. Also, Command data.

CNES – Centre National d'tudes Spatiales, France.

Conjunction – A configuration in which two celestial bodies have their least apparent separation.

Coherent – Two-way communications mode wherein the spacecraft generates its downlink frequency based upon the frequency of the uplink it receives.

Coma – The cloud of diffuse material surrounding the nucleus of a comet.

Comets – Small bodies composed of ice and rock in various orbits about the sun.

CRAF – Comet Rendezvous / Asteroid Flyby mission, cancelled.

CRS – Cosmic Ray Subsystem, high-energy particle instrument on Voyager.

CRT – Cathode ray tube video display device.

△

dB – Decibel, an expression of ratio (usually that of power levels) in the form of log base 10. A reference may be specified, for example, dBm is referenced to milliwatts, dBW is referenced to watts, etc. Example: 20 dBm = $10^{20/10}$ = 10^2 = 100 milliwatts.

DC – Direct current.

DC – The DSN Downlink Channel, several of which are in each DSN Downlink Tracking & Telemetry subsystem, DTT.

DCC – The DSN Downlink Channel Controller, one of which is in each DSN Downlink Channel, DC.

DCD – The DSN Data Capture and Delivery subsystem.

DCPC – The DSN Downlink Channel Processor Cabinet, one of which contains a DSN Downlink Channel, DC.

DEC – Declination.

Decibel – dB, an expression of ratio (see dB, above). One tenth of a Bel.

Declination – The measure of a celestial body's apparent height above or below the celestial equator.

Density – Mass per unit volume. For example, the density of water can be stated as 1 gram/cm^3.

Descending node – The point at which an orbit crosses a reference plane (such as a planet's equatorial plane or the ecliptic plane) going south.

DKF – DSN keyword file, also known as KWF.

Doppler effect – The effect on frequency imposed by relative motion between transmitter and receiver. See Chapter 6.

Downlink – Signal received from a spacecraft.

DSOT – Data System Operations Team, part of the DSMS staff.

DSCC – Deep Space Communications Complex, one of three DSN tracking sites at Goldstone, California; Madrid, Spain; and Canberra, Australia; spaced about equally around the Earth for continuous tracking of deep-space vehicles.

DSMS – Deep Space Mission System, the system of computers, software, networks, and procedures that processes data from the DSN at JPL.

DSN – Deep Space Network, NASA's worldwide spacecraft tracking facility managed and operated by JPL.

DSS – Deep Space Station, the antenna and front-end equipment at DSCCs.

DT – Dynamical Time. Replaces Ephemeris Time, ET, as the independent argument in dynamical theories and ephemerides. Its unit of duration is based on the orbital motions of the Earth, Moon, and planets. DT has two expressions, Terrestrial Time, TT, (or Terrestrial Dynamical Time, TDT), and Barycentric Dynamical Time, TDB. More information on these, and still more timekeeping expressions, may be found at the U.S. Naval Observatory website.

DTT – The DSN Downlink Tracking & Telemetry subsystem.

Dyne – A unit of force equal to the force required to accelerate a 1-g mass 1 cm per second per second. Compare with Newton.

△

E – East.

E – Exa, a multiplier, $\times 10^{18}$ from the Greek "hex" (six, the "h" is dropped). The reference to six is because this is the sixth multiplier in the series k, M, G, T, P, E. See the entry for CGPM.

Earth – Third planet from the sun, a terrestrial planet.

Eccentricity – The distance between the foci of an ellipse divided by the major axis.

Ecliptic – The plane in which Earth orbits the sun and in which solar and lunar eclipses occur.

EDL – (Atmospheric) Entry, Descent, and Landing.

EDR – Experiment Data Record.

EHz – ExaHertz (10^{18} Hz)

EL – Elevation.

Ellipse – A closed plane curve generated in such a way that the sums of its distances from the two fixed points (the foci) is constant.

ELV – Expendable launch vehicle.

EM – Electromagnetic.

EMF – Electromagnetic force (radiation).

EMR – Electromagnetic radiation.

EOT – End Of Track, used in DSN operations.

Equator – An imaginary circle around a body which is everywhere equidistant from the poles, defining the boundary between the northern and southern hemispheres.

Equinox – The equinoxes are times at which the center of the Sun is directly above the Earth's equator. The day and night would be of equal length at that time, if the Sun were a point and not a disc, and if there were no atmospheric refraction. Given the apparent disc of the Sun, and the Earth's atmospheric refraction, day and night actually become equal at a point within a few days of each equinox. The vernal equinox marks the beginning of spring in the northern hemisphere, and the autumnal equinox marks the beginning of autumn in the northern hemisphere.

ERC – NASA's Educator Resource Centers.

ERT – Earth-received time, UTC of an event at DSN receive-time, equal to SCET plus OWLT.

ESA – European Space Agency.

ESP – Extra-Solar Planet, a planet orbiting a star other than the Sun. See also Exoplanet.

ET – Ephemeris time, a measurement of time defined by orbital motions. Equates to Mean Solar Time corrected for irregularities in Earth's motions (obsolete, replaced by TT, Terrestrial Time).

eV – Electron volt, a measure of the energy of subatomic particles.

Exoplanet – Extrasolar planet. A planet orbiting a star other than the sun.

Extrasolar planet – A planet orbiting a star other than the sun. Exoplanet.

△

f, F – Force. Two commonly used units of force are the Newton and the dyne. Force = Mass × Acceleration.

FDS – Flight Data Subsystem.

FE – Far Encounter phase of mission operations.

Femtometer – 10^{-15} meter.

Fluorescence – The phenomenon of emitting light upon absorbing radiation of an invisible wavelength.

fm – Femtometer (10^{-15} m)

FM – Frequency modulation.

FTS – DSN Frequency and Timing System. Also, frequency and timing data.

FY – Fiscal year.

△

G – Universal Constant of Gravitation. Its tiny value (G = 6.6726 $\times 10^{-11}$ Nm^2/kg^2) is unchanging throughout the universe.

G – Giga, a multiplier, $\times 10^9$, from the Latin "gigas" (giant). See the entry for CGPM.

g – Acceleration due to a body's gravity. Constant at any given place, the value of g varies from object to object (e.g. planets), and also with the distance from the center of the object. The relationship between the two constants is: g = GM/r^2 where r is the radius of separation between the masses' centers, and M is the mass of the primary body (e.g. a planet). At Earth's surface, the value of g = 9.8 meters per second per second (9.8 m/s^2 or 9.8 ms^{-2}). See also weight.

g – Gram, a thousandth of the metric standard unit of mass (see kg). The gram was originally based upon the weight of a cubic centimeter of water, which still approximates the current value.

Gal – Unit of gravity field measurement corresponding to a gravitational acceleration of 1 cm/sec^2.

Galaxy – One of billions of systems, each composed of numerous stars, nebulae, and dust.

Galilean satellites – The four large satellites of Jupiter so named because Galileo discovered them when he turned his telescope toward Jupiter: Io, Europa, Ganymede, and Callisto.

Gamma rays – Electromagnetic radiation in the neighborhood of 100 femtometers wavelength.

GCF – Ground Communications Facilities, provides data and voice communications between JPL and the three DSCCs (obsolete, replaced by DCD).

GDS – Ground Data System, encompasses DSN, GCF, DSMS, and project data processing systems.

GDSCC – DSN's Goldstone Deep Space Communications Complex in California.

GEO – Geosynchronous Earth Orbit.

Geostationary – A geosynchronous equatorial circular orbit. Also called Clarke orbit.

Geosynchronous – A direct, circular, low inclination orbit about the Earth having a period of 23 hours 56 minutes 4 seconds.

GHz – Gigahertz (10^9 Hz).

GLL – The Galileo spacecraft.

GMT – Greenwich Mean Time (obsolete; UT, Universal Time is preferred).

Gravitation – The mutual attraction of all masses in the universe. Newton's Law of Universal Gravitation holds that every two bodies attract each other with a force that is directly proportional to the product of their masses, and inversely proportional to the square of the distance between them. This relation is given by the formula at right, where F is the force of attraction between the two objects, given G the Universal Constant of Gravitation, masses m1 and m2, and d distance. Also stated as $F_g = GMm/r^2$ where F_g is the force of gravitational attraction, M the larger of the two masses, m the smaller mass, and r the radius of separation of the centers of the masses. See also weight.

Gravitational waves – Einsteinian distortions of the space-time medium predicted by general relativity theory (not yet directly detected as of March 2011). (Not to be confused with gravity waves, see below.)

Gravity assist – Technique whereby a spacecraft takes angular momentum from a planet's solar orbit (or a satellite's orbit) to accelerate the spacecraft, or the reverse. See Chapter 4.

Gravity waves – Certain dynamical features in a planet's atmosphere (not to be confused with gravitational waves, see above).

Great circle – An imaginary circle on the surface of a sphere whose center is at the center of the sphere.

GSSR– Goldstone Solar System Radar, a technique which uses very high-power X-band and S-band transmitters at DSS 14 to illuminate solar system objects for imaging.

GTL – Geotail spacecraft.

GTO – Geostationary (or geosynchronous) Transfer Orbit.

△

HA – Hour Angle.

Halo orbit – A spacecraft's pattern of controlled drift about an unstable Lagrange point (L1 or L2 for example) while in orbit about the primary body (e.g. the Sun).

HEF – DSN's high-efficiency 34-m DSS, replaces STD DSSs.

Heliocentric – Sun-centered.

Heliopause – The boundary theorized to be roughly circular or teardrop-shaped, marking the edge of the sun's influence, perhaps 150 AU from the sun.

Heliosphere – The space within the boundary of the heliopause, containing the sun and solar system.

HEMT – High-electron-mobility transistor, a low-noise amplifier used in DSN.

HGA – High-Gain Antenna onboard a spacecraft.

Hohmann Transfer Orbit – Interplanetary trajectory using the least amount of propulsive energy. See Chapter 4.

h – Hour, 60 minutes of time.

Hour Angle – The angular distance of a celestial object measured westward along the celestial equator from the zenith crossing. In effect, HA represents the RA for a particular location and time of day.

△

ICE – International Cometary Explorer spacecraft.

ICRF – International Celestial Reference Frame. The realization of the ICRS provided by the adopted positions of extragalactic objects.

ICRS – International Celestial Reference System. Conceptual basis for celestial positions, aligned with respect to extremely distant objects and utilizing the theory of general relativity.

IERS – International Earth Rotation and Reference Systems Service.

IF – Intermediate Frequency. In a radio system, a selected processing frequency between RF (Radio Frequency) and the end product (e.g. audio frequency).

Inclination – The angular distance of the orbital plane from the plane of the planet's equator, stated in degrees.

IND – JPL's Interplanetary Network Directorate, formerly IPN-ISD.

Inferior planet – Planet which orbits closer to the Sun than the Earth's orbit.

Inferior conjunction – Alignment of Earth, sun, and an inferior planet on the same side of the sun.

Ion – A charged particle consisting of an atom stripped of one or more of its electrons.

IPAC – Infrared Processing and Analysis Center at Caltech campus on Wilson Avenue in Pasadena.

IPC – Information Processing Center, JPL's computing center on Woodbury Avenue in Pasadena.

IPN-ISD – JPL's Interplanetary Network and Information Systems Directorate, formerly TMOD (obsolete. See IND).

IR – Infrared, meaning "below red" radiation. Electromagnetic radiation in the neighborhood of 100 micrometers wavelength.

IRAS – Infrared Astronomical Satellite.

ISM – Interstellar medium.

ISO – International Standards Organization.

ISOE – Integrated Sequence of Events.

Isotropic – Having uniform properties in all directions.

IUS – Inertial Upper Stage.

△

JGR – Journal Of Geophysical Research.

Jovian – Jupiter-like planets, the gas giants Jupiter, Saturn, Uranus, and Neptune.

JPL – Jet Propulsion Laboratory, operating division of the California Institute of Technology.

Jupiter – Fifth planet from the sun, a gas giant or Jovian planet.

△

k – Kilo, a multiplier, $\times 10^3$ from the Greek "khilioi" (thousand). See the entry for CGPM.

K – Kelvin, the SI base unit of thermodynamic temperature.

K-band – A range of microwave radio frequencies in the neighborhood of 12 to 40 GHz.

kg – Kilogram. See below.

Keyhole – An area in the sky where an antenna cannot track a spacecraft because the required angular rates would be too high. Mechanical limitations may also contribute to keyhole size. Discussed in depth under Chapter 2.

kHz – kilohertz.

Kilogram (kg) – the SI base unit of mass, based on the mass of a metal cylinder kept in France. See also g (gram).

Kilometer – 10^3 meter.

Klystron – A microwave travelling wave tube power amplifier used in transmitters.

km – Kilometers.

KSC – Kennedy Space Center, Cape Canaveral, Florida.

KWF – Keyword file of events listing DSN station activity. Also known as DKF, DSN keyword file.

Kuiper belt – A disk-shaped region about 30 to 100 AU from the sun considered to be the source of the short-period comets.

△

Lagrange points – Five points with respect to an orbit which a body can stably occupy. Designated L1 through L5. See Chapter 5.

LAN – Local area network for inter-computer communications.

Large Magellanic Cloud – LMC, the larger of two small galaxies orbiting nearby our Milky Way galaxy, which are visible from the southern hemisphere.

Laser – Light Amplification by Stimulated Emission of Radiation. Compare with Maser.

Latitude – Circles in parallel planes to that of the equator defining north-south measurements, also called parallels.

L-band – A range of microwave radio frequencies in the neighborhood of 1 to 2 GHz.

LCP – Left-hand circular polarization.

Leap Second – A second which may be added or subtracted to adjust UTC at either, both, or neither, of two specific opportunities each year.

Leap Year – Every fourth year, in which a 366th day is added since the Earth's revolution takes 365 days 5 hr 49 min.

LECP – Low-Energy Charged-Particular Detector onboard a spacecraft.

LEO – Low Equatorial Orbit.

LGA – Low-Gain Antenna onboard a spacecraft.

Light – Electromagnetic radiation in the neighborhood of 1 nanometer wavelength.

Light speed – 299,792 km per second, the constant c.

Light time – The amount of time it takes light or radio signals to travel a certain distance at light speed.

Light year – A measure of distance, the distance light travels in one year, about 63,197 AU.

LMC – Large Magellanic Cloud, the larger of two small galaxies orbiting nearby our Milky Way galaxy, which are visible from the southern hemisphere.

LMC – Link Monitor and Control subsystem at the SPCs within the DSN DSCCs.

LNA – Low-noise amplifier in DSN, either a maser or a HEMT.

Local time – Time adjusted for location around the Earth or other planets in time zones.

Longitude – Great circles that pass through both the north and south poles, also called meridians.

LOS – Loss Of Signal, used in DSN operations.

LOX – Liquid oxygen.

△

m – Meter (U.S. spelling; elsewhere metre), the international standard of linear measurement.

m – milli- multiplier of one one-thousandth, e.g. 1 mW = 1/1000 of a watt, mm = 1/1000 meter.

m, M – Mass. The kilogram is the standard unit of mass. Mass = Acceleration / Force.

M – Mega, a multiplier, $\times 10^6$ (million) from the Greek "megas" (great). See the entry for CGPM.

M100 – Messier Catalog entry number 100 is a spiral galaxy in the Virgo cluster seen face-on from our solar system.

Major axis – The maximum diameter of an ellipse.

Mars – Fourth planet from the sun, a terrestrial planet.

Maser – A microwave travelling wave tube amplifier named for its process of Microwave Amplification by Stimulated Emission of Radiation. Compare with Laser. In the Deep Space Network, masers are used as low-noise amplifiers of downlink signals, and also as frequency standards.

Mass – A fundamental property of an object comprising a numerical measure of its inertia; the amount of matter in the object. While an object's mass is constant (ignoring Relativity for this purpose), its weight will vary depending on its location. Mass can only be measured in conjunction with force and acceleration.

MC-cubed or **MC³** – Mission Control and Computing Center at JPL (outdated).

MCCC – Mission Control and Computing Center at JPL (outdated).

MCD – DSN's maximum-likelyhood convolutional decoder, the Viterbi decoder.

MCT – Mission Control Team, JPL Section 368 mission execution real-time operations.

MDSCC – DSN's Madrid Deep Space Communications Complex in Spain.

Mean solar time – Time based on an average of the variations caused by Earth's non-circular orbit. The 24-hour day is based on mean solar time.

Mercury – First planet from the sun, a terrestrial planet.

Meridians – Great circles that pass through both the north and south poles, also called lines of longitude.

MESUR – The Mars Environmental Survey project at JPL, the engineering prototype of which was originally called MESUR Pathfinder, later Mars Pathfinder.

Meteor – A meteoroid which is in the process of entering Earth's atmosphere. It is called a meteorite after landing.

Meteorite – Rocky or metallic material which has fallen to Earth or to another planet.

Meteoroid – Small bodies in orbit about the sun which are candidates for falling to Earth or to another planet.

MGA – Medium-Gain Antenna onboard a spacecraft.

MGN – The Magellan spacecraft.

MGSO –JPL's Multimission Ground Systems Office (obsolete; see TMOD, IND).

MHz – Megahertz (10^6 Hz).

Micrometer – μm, or μ, 10^{-6} meter.

Micron – Obsolete term for micrometer, μm (10^{-6} m).

Milky Way – The galaxy which includes the Sun and Earth.

Millimeter – 10^{-3} meter.

MIT – Massachusetts Institute of Technology.

MLI – Multi-layer insulation (spacecraft blanketing). See Chapter 11.

mm – millimeter (10^{-3} m).

MO – The Mars Observer spacecraft.

Modulation – The process of modifying a radio frequency by shifting its phase, frequency, or amplitude to carry information.

MON – DSN Monitor System. Also, monitor data.

Moon – A small natural body which orbits a larger one. A natural satellite. Capitalized, the Earth's natural satellite.

MOSO – Multimission Operations Systems Office at JPL.

MR – Mars relay.

μm – Micrometer (10^{-6} m).

μ – Micrometer (10^{-6} m).

Multiplexing – A scheme for delivering many different measurements in one data stream. See Chapter 10.

$\triangle$

N – Newton, the SI unit of force equal to that required to accelerate a 1-kg mass 1 m per second per second (1m/sec^2). Compare with dyne.

N – North.

Nadir – The direction from a spacecraft directly down toward the center of a planet. Opposite the zenith.

NASA – National Aeronautics and Space Administration.

NE – Near Encounter phase in flyby mission operations.

Neptune – Eighth planet from the sun, a gas giant or Jovian planet.

NiCad – Nickel-cadmium rechargable battery.

NIMS – Near-Infrared Mapping Spectrometer onboard the Galileo spacecraft.

NIST – National Institute of Standards.

nm – Nanometer (10^{-9} m).

nm – Nautical Mile, equal to the distance spanned by one minute of arc in latitude, 1.852 km.

NMC – Network Monitor and Control subsystem in DSN.

NOCC – DSN Network Operations Control Center at JPL.

Nodes – Points where an orbit crosses a reference plane.

Non-coherent – Communications mode wherein a spacecraft generates its downlink frequency independent of any uplink frequency.

Nucleus The central body of a comet.

Nutation – A small nodding motion in a rotating body. Earth's nutation has a period of 18.6 years and an amplitude of 9.2 arc seconds.

NRZ – Non-return to zero. Modulation scheme in which a phase deviation is held for a period of time in order to represent a data symbol. See Chapter 10.

NSP – DSN Network Simplification Project. A project that re-engineered the DSN to consolidate seven data systems into two data systems that handle the same data types.

$\triangle$

OB – Observatory phase in flyby mission operations encounter period.

One-way – Communications mode consisting only of downlink received from a spacecraft.

Oort cloud – A large number of comets theorized to orbit the sun in the neighborhood of 50,000 AU.

OPCT – Operations Planning and Control Team at JPL, "OPSCON" (bsolete, replaced by DSOT, Data Systems Operations Team).

Opposition – Configuration in which one celestial body is opposite another in the sky. A planet is in opposition when it is 180 degrees away from the sun as viewed from another planet (such as Earth). For example, Saturn is at opposition when it is directly overhead at midnight on Earth.

OPNAV – Optical Navigation (images).

OSI – ISO's Open Systems Interconnection protocol suite.

OSR – Optical Solar Reflector, thermal control component onboard a spacecraft.

OSS – Office Of Space Science, NASA (obsolete, replaced by Science Mission Directorate SMD).

OSSA – Office Of Space Science and Applications, NASA (obsolete, see OSS).

OTM – Orbit Trim Maneuver, spacecraft propulsive maneuver.

OWLT – One-Way Light Time, elapsed time between Earth and spacecraft or solar system body.

△

P – Peta, a multiplier, $\times 10^{15}$, from the Greek "pente" (five, the "n" is dropped). The reference to five is because this is the fifth multiplier in the series k, M, G, T, P. See the entry for CGPM.

Packet – A quantity of data used as the basis for multiplexing, for example in accordance with CCSDS.

PAM – Payload Assist Module upper stage.

Parallels – Circles in parallel planes to that of the equator defining north-south measurements, also called lines of latitude.

Pathfinder – The Mars Environmental Survey (MESUR) engineering prototype later named Mars Pathfinder.

PDS – Planetary Data System.

PDT – Pacific Daylight Time.

PE – Post Encounter phase in flyby mission operations.

Periapsis – The point in an orbit closest to the body being orbited.

Perigee – Periapsis in Earth orbit.

Perichron – Periapsis in Saturn orbit.

Perihelion – Periapsis in solar orbit.

Perijove – Periapsis in Jupiter orbit.

Perilune – Periapsis in lunar orbit.

Periselene – Periapsis in lunar orbit.

Phase – The angular distance between peaks or troughs of two waveforms of similar frequency.

Phase – The particular appearance of a body's state of illumination, such as the full or crescent phases of the Moon.

Phase – Any one of several predefined periods in a mission or other activity.

Photovoltaic – Materials that convert light into electric current.

PHz – Petahertz (10^{15} Hz).

PI – Principal Investigator, scientist in charge of an experiment.

Picometer – 10^{-12} meter.

PIO– JPL's Public Information Office.

Plasma – Electrically conductive fourth state of matter (other than solid, liquid, or gas), consisting of ions and electrons.

PLL – Phase-lock-loop circuitry in telecommunications technology.

Plunge – In describing the tracking motion of an AZ-EL or ALT-AZ mounted radio telescope, to "plunge" means to exceed 90 in elevation and then to continue tracking as elevation decreases on the other side without swiveling around in azimuth. This is not a capability of DSN antennas.

Pluto – Dwarf planet, Kuiper Belt Object, and Trans-Neptunian Object.

pm – Picometer (10^{-12} m).

PM – Post meridiem (Latin: after midday), afternoon.

PN10 – Pioneer 10 spacecraft.

PN11 – Pioneer 11 spacecraft.

Prograde – Orbit in which the spacecraft moves in the same direction as the planet rotates. See retrograde.

PST – Pacific Standard Time.

PSU – Pyrotechnic Switching Unit onboard a spacecraft.

△

Quasar – Quasi-stellar object observed mainly in radio waves. Quasars are extra-galactic objects believed to be the very distant centers of active galaxies.

△

RA – Right Ascension.

Radian – Unit of angular measurement equal to the angle at the center of a circle subtended by an arc equal in length to the radius. Equals about 57.296 degrees.

RAM – Random Access Memory.

RCP – Right-hand circular polarization.

Red dwarf – A small star, on the order of 100 times the mass of Jupiter.

Reflection – The deflection or bouncing of electromagnetic waves when they encounter a surface.

Refraction – The deflection or bending of electromagnetic waves when they pass from one kind of transparent medium into another.

REM – Receiver Equipment Monitor within the Downlink Channel (DC) of the Downlink Tracking & Telemetry subsystem (DTT).

Retrograde – Orbit in which the spacecraft moves in the opposite direction from the planet's rotatation. See prograde.

RF – Radio Frequency.

RFI – Radio Frequency Interference.

Right Ascension – The angular distance of a celestial object measured in hours, minutes, and seconds along the celestial equator eastward from the vernal equinox.

Rise – Ascent above the horizon.

RNS – GCF reliable network service (obsolete, replaced by DCD).

ROM – Read Only Memory.

RPIF – Regional Planetary Imaging Data Facilities.

RPS – Radioisotope Power System.

RRP – DSN Receiver & Ranging Processor within the Downlink Channel (DC) of the Downlink Tracking & Telemetry subsystem (DTT).

RS – DSN Radio Science System. Also, radio science data.

RTG – Radioisotope Thermo-Electric Generator onboard a spacecraft.

RTLT – Round-Trip Light Time, elapsed time roughly equal to 2× OWLT.

△

S – South.

s – Second, the SI base unit of time.

SA – Solar Array, photovoltaic panels onboard a spacecraft.

SAF – Spacecraft Assembly Facility, JPL Building 179.

SAR – Synthetic Aperture Radar

Satellite – A small body which orbits a larger one. A natural or an artificial moon. Earth-orbiting spacecraft are called satellites. While deep-space vehicles are technically satellites of the sun or of another planet, or of the galactic center, they are generally called spacecraft instead of satellites.

Saturn – Sixth planet from the sun, a gas giant or Jovian planet.

S-band – A range of microwave radio frequencies in the neighborhood of 2 to 4 GHz.

SC – Steering Committee.

SCET – Spacecraft Event Time, equal to ERT minus OWLT.

SCLK – Spacecraft Clock Time, a counter onboard a spacecraft.

Sec – Abbreviation for Second.

Second – the SI base unit of time.

SEDR – Supplementary Experiment Data Record.

SEF – Spacecraft event file.

SEGS – Sequence of Events Generation Subsystem.

Semi-major axis – Half the distance of an ellipse's maximum diameter, the distance from the center of the ellipse to one end.

Set – Descent below the horizon.

SFOF – Space Flight Operations Facility, Buildings 230 and 264 at JPL.

SFOS – Space Flight Operations Schedule, product of SEGS.

Shepherd moons – Moons which gravitationally confine ring particles.

SI – The International System of Units (metric system). See also Units of Measure.

SI base unit – One of seven SI units of measure from which all the other SI units are derived. See SI derived unit. See also Units of Measure.

SI derived unit – One of many SI units of measure expressed as relationships of the SI base units. For example, the watt, W, is the SI derived unit of power. It is equal to joules per second. $W = m^2 \cdot kg \cdot s^3$ (Note: the joule, J, is the SI derived unit for energy, work, or quantity of heat.) See also Units of Measure.

Sidereal time – Time relative to the stars other than the sun.

SIRTF – Space Infrared Telescope Facility.

SMC – Small Magellanic Cloud, the smaller of two small galaxies orbiting nearby our Milky Way galaxy, which are visible from the southern hemisphere.

SMD – Science Mission Directorate, NASA (previously Office Of Space Science, OSS).

SOE – Sequence of Events.

Solar wind – Flow of lightweight ions and electrons (which together comprise plasma) thrown from the sun.

SNR – Signal-to-Noise Ratio.

SPC – Signal Processing Center at each DSCC.

Specific Impulse – A measurement of a rocket's relative performance. Expressed in seconds, the number of which a rocket can produce one pound of thrust from one pound of fuel. The higher the specific impulse, the less fuel required to produce a given amount of thrust.

Spectrum – A range of frequencies or wavelengths.

SSA – Solid State Amplifier in a spacecraft telecommunications subsystem, the final stage of amplification for downlink.

SSI – Solid State Imaging Subsystem, the CCD-based cameras on Galileo.

SSI – Space Services, Inc., Houston, manufacturers of the Conestoga launch vehicle.

STD – Standard 34-m DSS, retired from DSN service.

STS – Space Transportation System (Space Shuttle).

Subcarrier – Modulation applied to a carrier which is itself modulated with information-carrying variations.

Sun synchronous orbit – A spacecraft orbit that precesses, wherein the location of periapsis changes with respect to the planet's surface so as to keep the periapsis location near the same local time on the planet each orbit. See walking orbit.

Superior planet – Planet which orbits farther from the sun than Earth's orbit.

Superior conjunction – Alignment between Earth and a planet on the far side of the sun.

SWG – Science Working Group.

△

TAU – Thousand AU Mission.

TCM – Trajectory Correction Maneuver, spacecraft propulsive maneuver.

TDM – Time-division multiplexing.

Termination shock – Shock at which the solar wind is thought to slow to subsonic speed, well inside the heleopause.

T – Tera, a multiplier $\times 10^{12}$, from the Greek teras (monster). See the entry for CGPM.

Terrestrial planet – One of the four inner Earth-like planets.

Three-way – Coherent communications mode wherein a DSS receives a downlink whose frequency is based upon the frequency of an uplink provided by another DSS.

TMOD – (obsolete. See IPN-ISD) JPL's Telecommunications and Mission Operations Directorate. Formerly MGSO.

THz – Terahertz (10^{12} Hz).

TLM – DSN Telemetry data.

TLP – DSN Telemetry Processor within the DTT Downlink Channel.

TOS – Transfer Orbit Stage, upper stage.

Transducer – Device for changing one kind of energy into another, typically from heat, position, or pressure into a varying electrical voltage or vice-versa, such as a microphone or speaker.

Transponder – Electronic device which combines a transmitter and a receiver.

TRC – NASA's Teacher Resource Centers (obsolete, now called Educator Resource Centers, ERC).

TRK – DSN Tracking System. Also, Tracking data.

TRM – Transmission Time, UTC Earth time of uplink.

True anomaly – The angular distance of a point in an orbit past the point of periapsis, measured in degrees.

TWNC – Two-Way Non-Coherent mode, in which a spacecraft's downlink is not based upon a received uplink from DSN.

Two-way – Communications mode consisting of downlink received from a spacecraft while uplink is being received at the spacecraft. See also coherent.

TWT – Traveling Wave Tube, downlink power amplifier in a spacecraft telecommunications subsystem, the final stage of amplification for downlink (same unit as TWTA).

TWTA – Traveling Wave Tube Amplifier, downlink power amplifier in a spacecraft telecommunications subsystem, the final stage of amplification for downlink (same unit as TWT).

TXR – DSN's DSCC Transmitter assembly.

△

UHF – Ultra-high frequency (around 300 MHz).

μm – Micrometer (10^{-6} m).

ULS – Ulysses spacecraft.

Uplink – Signal sent to a spacecraft.

UPL – The DSN Uplink Tracking & Command subsystem.

Uranus – Seventh planet from the sun, a gas giant or Jovian planet.

USO – Ultra Stable Oscillator, in a spacecraft telecommunications subsystem.

UT – Universal Time, also called Zulu (Z) time, previously Greenwich Mean Time. UT is based on the imaginary "mean sun," which averages out the effects on the length of the solar day caused by Earth's slightly non-circular orbit about the sun. UT is not updated with leap seconds as is UTC.

UTC – Coordinated Universal Time, the world-wide scientific standard of time-keeping. It is based upon carefully maintained atomic clocks and is highly stable. Its rate does not change by more than about 100 picoseconds per day. The addition or subtraction of leap seconds, as necessary, at two opportunities every year adjusts UTC for irregularities in Earth's rotation. The U.S. Naval Observatory website provides information in depth on the derivation of UTC.

UV – Ultraviolet (meaning "above violet") radiation. Electromagnetic radiation in the neighborhood of 100 nanometers wavelength.

UWV – DSN Microwave subsystem in DSSs which includes waveguides, waveguide switches, LNAs, polarization filters, etc.

△

Velocity – A vector quantity whose magnitude is a body's speed and whose direction is the body's direction of motion.

Venus – Second planet from the sun, a terrestrial planet.

VGR1 – Voyager 1 spacecraft.

VGR2 – Voyager 2 spacecraft.

VLBI – DSN Very Long Baseline Interferometry System. Also, VLBI data.

△

W – Watt, a measure of electrical power equal to potential in volts times current in amps.

W – West.

Walking orbit – A spacecraft orbit that precesses, wherein the location of periapsis changes with respect to the planet's surface in a useful way. See sun-synchronous.

Wavelength – The distance that a wave from a single oscillation of electromagnetic radiation will propagate during the time required for one oscillation.

Weight – The gravitational force exerted on an object of a certain mass. The weight of mass m is mg Newtons, where g is the local acceleration due to a body's gravity.

WWW – World-Wide Web.

△

X-band – A range of microwave radio frequencies in the neighborhood of 8 to 12 GHz.

X-ray – Electromagnetic radiation in the neighborhood of 100 picometer wavelength.

△

Y – Yotta, a multiplier, $\times 10^{24}$ from the second-to-last letter of the Latin alphabet. See the entry for CGPM.

△

Z – Zetta, a multiplier, $\times 10^{21}$ from the last letter of the Latin alphabet. See the entry for CGPM.

Z – Zulu in phonetic alphabet, stands for UT, Universal Time.

Zenith – The point on the celestial sphere directly above the observer. Opposite the nadir.

Index

This book is set in the font Computer Modern–
Super (2008), designed by Donald Knuth.